SCIENTIFIC BLUNDERS

SCIENTIFIC BLUNDERS

A brief history
of how wrong scientists can sometimes be . . .

ROBERT M. YOUNGSON

Carroll & Graf Publishers, Inc.
NEW YORK

Carroll & Graf Publishers, Inc.
19 West 21st Street
New York
NY 10010–6805

First published in the UK by Robinson Publishing Ltd 1998

First Carroll & Graf edition 1998

Copyright © Robert M. Youngson 1998

ISBN 0–7867–0594–9

Printed and bound in the United Kingdom

10 9 8 7 6 5 4 3 2 1

Contents

Contents

Contents

To Eric Cornes

Without whose unfailing interest
and many helpful suggestions
this book would have been finished in half the time

Preface

The history of science is littered with remarkable errors, many of which are, of course, no more than aspects of the scientific ignorance of the time. Even so, they are among the most momentous errors in the history of humankind and many of them have had serious consequences. This makes them particularly interesting and instructive. Some blunders are the result of carelessness; some arise from plain, stubborn wrong-headedness; some from arrogance (as we shall see, *hubris* has always been common in the upper echelons of the scientific establishment); some arise from wilful and culpable ignorance; some from spectacular bad luck; and some from human moral frailty. Nevertheless, however incredible the bald facts of each case may seem, there are nearly always extenuating factors, and, in spite of our natural tendency to look for the worst in other people, it is good for us to remember this and to look tolerantly for these mitigating circumstances.

Other people's errors and blunders are strangely fascinating and there is no reason why we should not enjoy a little harmless gratification at other people's expense, but there are better motives for reading about blunders than the expectation of feeling superior. By investigating the circumstances in which blunders occur, we can learn a great deal, not only about human nature but also about the subject that is the occasion for the blunder. This, as you will find if you ever get around to reading this book, is particularly so in the case of scientific blunders.

Paradoxically, for the general non-scientific reader, this is probably the best way of learning about science – as a review of most of the stories in this book will quickly show. Getting it wrong is very often the way that science advances. Provisional, but wrong, ideas give way to better – but still wrong – ideas. Every stage in this process may involve a blunder. However, as theories account more and more fully for the observed facts, the blunder factor is gradually attenuated. It is, of course, never eliminated. For the non-professional, an account of this stage-by-stage process can be an almost painless way of getting a real grasp of current scientific principles. And, believe me, if this happens to you, it will change your life.

A study of blunders can even benefit the professional scientist. This point was brought out by the great philosopher of science Karl Popper (1902–94), who commented that scientific blunders actually add to knowledge by demonstrating what is wrong with a hypothesis (a proposed explanation of something). Here is a case in point. The physiologist John Carew Eccles (1903–97) was convinced that the mind was independent of the brain. He allowed this conviction – which is now universally dismissed by scientists – to influence his work on nerve transmission. As a result he spent years on research that was based on a completely wrong hypothesis. Needless to say, this work came to nothing. Discouraged, Eccles met Popper, who persuaded him that backing the wrong hypothesis had not been a failure but a success, because it showed that the truth lay elsewhere. With renewed enthusiasm, Eccles abandoned his former theories and started work on the function of synapses that eventually led to enormously important advances in medicine and pharmacology. It also, incidentally, earned him a Nobel Prize and a knighthood.

Many scientific blunders happen because the person concerned is more preoccupied with the possible rewards of a discovery – fame, status, professional advancement, perhaps even wealth – than with the disinterested pursuit of truth. When this happens, there is danger of bias in favour of the hoped-for outcome. Leaning in the desired direction can cause one of the most serious kinds of scientific blunder – to be tempted into dishonesty. It is very hard for the scientist to maintain his or her moral integrity at the expense of months or years of promising work. As Thomas Henry Huxley put it: 'The great tragedy of science is the slaying of a beautiful hypothesis by an ugly fact.'

The scientist who finds himself, or herself, ignoring – or explaining away – uncomfortable results that don't fit in with the preconceived hypothesis is on a slippery slope and is liable to end up in disgrace and ruin. Most scientific workers are aware of their fallibility and take precautions to avoid the blunder of bias in the interpretation of results. The best of them are said always see to it that if findings are susceptible to such bias it always acts *against* the favoured hypothesis. I may be cynical, but I take leave to doubt that such angelic behaviour can be very common. Scientists are just people and are not exempt from the human weaknesses that beset us all. Plain unvarnished deliberate fraud cannot, of course, be classed as a blunder, but there must be very few scientists who set out with the deliberate intention of perpetrating a scientific deception. They are all much too intelligent to think that they could ever get away with it.

Some of the most breathtaking scientific blunders have been the work of non-scientists – groups motivated by political, religious or other ideologies. The influence of such groups in opposing demonstrable truth can be enormously damaging. These groups have their own versions of truth and their truths are important to them. It is when their paradigms conflict with those of science that the trouble arises. This conflict has been a major feature of the history of human thought and is by no means resolved today. Few things can be more important than to try to reconcile these fundamental differences. I hope, perhaps quixotically, that by looking at the reasons for some of these conflicts, a little light may be shed on them.

Science is much too important to be brushed aside. To quote one of the wisest and most balanced scientists of modern times, Nobel Laureate Peter Medawar (1915–87): 'Science and technology are incomparably the most successful endeavours human beings have ever engaged upon.' The force of this view is not a whit diminished by the critics of science, however vociferous, because attitudes to science are determined almost entirely by how much people know about it. The most noisy opponents are invariably the most ignorant of science. I hope that by the time you have finished reading this book you will know enough about the subject to be able to share my view of this and agree.

Unfortunately, for people not professionally engaged in it, science is undoubtedly difficult and it is not easy to gain an adequate knowledge of it. A straight account of scientific principles, however popularly expressed, is unlikely to succeed. In such a case it is necessary to take a roundabout route, and that is why I have written this book.

Robert M. Youngson
Blandford Forum, 1998

Cosmology and Earth Science

> The doctrine that the earth is not the centre of the universe, that it is not immovable but moves with a daily rotation, is absurd and both philosophically and theologically false.
> *The Congregation of the Inquisition against Galileo*

Star errors

The trouble with the word 'cosmology' is that it means different things to different people. To the philosopher and the theologian, cosmology is a speculative study of the meaning of the universe. To the historian of science, the term refers to a particular account of the universe as dreamed up by various imaginative people in the past. To the astronomer, it means the study of what we can find out about the structure of the universe. To the astrophysicist, cosmology is concerned with the theories of the origins and evolution of the universe.

There are not too many professional cosmologists, and scientific cosmologists – as distinct from philosophers, theologians and historians – have to know a lot. As well as covering the whole of astronomy, they also need a thorough grasp of mathematics and physics. But in a sense we are all cosmologists. From the earliest times people have wondered about the origin and nature of the universe, and since people first put down their thoughts in writing, many of these speculations have been documented and preserved.

Before we get to the stage of scientific blundering, we need a brief review of what went before. There were plenty of blunders, but they were not scientific. Early cosmology came before science and had its origins in the age of magic when the best explanation people could come up with was that the world was run by ghosts or spirits – sometimes friendly, but more often evil and malign. These ideas were necessarily based on human experience, and the spirits were credited with the same emotions and motivation as those of people but, of course, with a great deal more power. Even today, none of us is capable of envisaging anything totally different from what we know: we are the prisoners of our experience. This is why the word 'anthropomorphic' (human-shaped) is so often used in connection with accounts of this kind. It also explains why any attempt to account for matters outside our experience will usually produce statements that may sound significant but are actually meaningless.

Like most things on earth, the sun, moon and stars were the abode of spirits, and all natural phenomena were attributed to the ill-will, or approval, of these spirits. In *The Golden Bough* (1890), the Scottish anthropologist James Frazer (1854–1941) describes how, as knowledge of the world grew, magic gradually evolved into myth and religion. Early people shared our need for explanations and their writings are full of accounts of creation myths. The earliest known are those of the Sumerians, who lived in the fertile region between the great Tigris and Euphrates rivers, the area occupied by modern-day Iraq. They decided that everything began in an 'encircling watery abyss' which was acted on by a blind force to produce the gods and goddesses that created the universe. The ancient Egyptian myths also involve a watery abyss in which lived a formless spirit that carried within it the source of all existence, which created the gods and goddesses and the universe.

In fact, these statements don't actually explain anything. Oddly enough, people's desire for 'explanations' are readily satisfied by firm statements, especially if delivered in an authoritative tone, even if they don't mean a thing. This characteristic of humans has not changed since the time of the Sumerians, six thousand years ago. You can come across examples of this every day, and not only in the case of simple-minded and ignorant people. One of the responsibilities of science is to distinguish between meaningful statements and merely emotionally gratifying noises.

The Indian myths did rather better than those of the Sumerians. Although they contained an immensely complicated history of gods, they were more sophisticated in acknowledging that the ultimate origin must remain unknown. The Rig-Veda asks: 'Who can speak of the origins of creation? Did he who controls this world make it? Does he know?' Later documents, commenting on the beginning, state: 'All was darkness, without form, beyond reason and perception, like sleep.' From this arose the all-creating Lord.

Chinese ideas began almost to border on science. They were complex and subtle and incorporated the concept that creation was based on earth, air, fire and water – a completely wrong idea which was also later adopted by the Greeks – but, at least, an attempt at a unifying explanation. They also incorporated the principle of the opposing but complementary qualities of yin and yang. This is an elegant concept, but there is no objective evidence that it is anything more than that. The early Greeks decided that the universe was made from four things, or, in more impressive language, primordial entities – the void or Chaos; the earth or Gaea; the lower world, known as Tartarus; and love or Eros. These somehow worked together to give rise to a whole collection, or pantheon, of gods, including Uranus, the sky god. In the Christian tradition, creation, as recounted in the book of Genesis, was the work of a single God.

All early cosmological ideas quite naturally put humans at the centre of the universe. Aristotle's cosmology of the fourth century BC placed the earth at the middle of the universe, and this idea, which was also held by the second century Greek astronomer Claudius Ptolemaeus (*c*.90–168), better known as Ptolemy, actually persisted until the beginning of the sixteenth century. Ptolemy's great compendium of astronomy, the *Almagest*, contained a star map and a treatise on the fixed stars; this was based on the work of an earlier Greek astronomer called Hipparchus (*fl.* second century BC), who had set forth the basic principle of astronomy and compiled a catalogue of same 850 fixed stars. Ptolemy was less interested in science than in putting across his cosmological ideas. The fact is that the Greeks were not very keen on real experimental science, which they considered to be beneath their dignity. They considered that knowledge should be obtained by pure thought. Philosophy was the proper activity of a gentleman; finding out by looking and trying was strictly for slaves and other inferiors. This was a major blunder.

Simply on the grounds of 'a sense of the fitness of things', Ptolemy assumed that the earth was the central pivot around which the sun and stars revolved. Surrounding the earth, he stated, were eight concentric transparent spheres that could rotate around each other. Each of the first seven carried a single heavenly body – the moon, sun and five planets – while the eighth carried all the fixed stars. Beyond were the heavens, the abode of the blessed. Ptolemy was perfectly aware that there were unequivocal observations that did not fit into this scheme, but he was more interested in expounding his imaginary theories than in accounting for undeniable facts. So he had to produce some wildly elaborate explanations to account for such things as the apparent irregularity of the movement of the planets.

So far, we are at a fairly primitive stage of science – the stage of bold assertion without proof. Ptolemy's speculations were pure imagination and were not based on observation. No one knows when the first serious astronomical observations were made. Some people think they go back to the times when the ancient stone circles and other megalithic structures were set up. Stonehenge, for instance, which probably dates from around 3000 BC, may possibly have been intended as some kind of astronomical instrument. It is impossible to be sure about this. We have to come to much more recent times to see the dawning of true scientific observation.

Much reliable astronomical information was derived from naked eye observation, long before the telescope was invented. This observation was not always as careful as it might have been and, for instance, many tables of the movements of the planets, relative to the fixed stars, were published with misleading data. Some of the early observations were, however, remarkable, notably those of the great Polish astronomer Nicolas Copernik (1473–1543), better known as Copernicus, the latinized form which it was then fashionable to use, who, incidentally, spent many years correcting old tables. Among many other things, Copernicus was actually able, simply by careful observation and record-keeping, to show that if you imagine the earth's axis extended out into space, it will be found to move in a circle and will eventually generate a cone. This was the explanation of the slow movement of the position in the year of the two points at which the day and night are equal (the precession of the equinoxes). This phenomenon had in fact been described by the above-mentioned Greek astronomer Hipparchus some sixteen hundred years before.

After naked eye observation came the dawn of scientific astronomical instruments. The first important one was the quadrant – a simple form of protractor with a sighting rule, or alidade, used to measure the angle between a star, or the sun, and the horizontal. A plumb line was used to ensure that the instrument was kept horizontal. With a quadrant, you could measure the altitude of the North Star and then, by checking the time from the altitude of the sun and consulting a table, establish the latitude of the point from which the observation was made. This was a matter of some importance to ocean-going stalwarts.

The astrolabe – a term meaning 'star finder' – was a more refined form of quadrant and was made in various models. One popular and portable form consisted of two flat metal discs, 7.5 to 25 centimetres (3 to 10 inches) in diameter, capable of being rotated on a common centre. One side of one disc had degrees around its edge and was equipped with an alidade for sighting the star or the sun. The other side was engraved with a star map and lines to show the horizon, the zenith and the lines of azimuth and altitude for particular latitudes. The most important treatise on the astrolabe – a book that was studied for centuries by navigators and astronomers – was, surprisingly, written by the English poet and humorist Geoffrey Chaucer (1340–1400) of Canterbury Pilgrims fame. In the eighteenth century, the astrolabe was replaced by the more accurate sextant, invented by Isaac Newton.

The most notable earliest exponent of the quadrant was the Danish pioneer of astronomy Tycho Brahe (1546–1601) whose work marked the real beginning of scientific astronomical research. Tycho did not have the benefit of a telescope, which had not yet been invented, and had to develop new instruments for himself. Among other things, he designed and had constructed an enormous but rather crude astronomical quadrant of radius fourteen cubits (about 7 metres), divided into minutes. This quadrant enabled him to locate the position of hundreds of fixed stars with remarkable accuracy, and made him famous.

Tycho's enthusiasm for science and insistence on exactness paved the way for future advances and inspired his assistant Johannes Kepler to even greater efforts. In November 1572 Tycho observed a new star in the constellation Cassiopeia, where no star had previously been observed. This star was brighter than Venus but faded away in 17 months. Tycho's observations recorded all its changes, and proved that it lay beyond the moon at an immeasurable distance away. He published an account of this in the book *De nova stella* in 1573 and soon had an international reputation as an astronomer. Since that time, such exploding stars have been called 'novas'.

Tycho was one of those characters, unusual but immensely important in the history of science, whose contribution consisted less in the advancement of knowledge than in the promotion of method and in the stimulation of enthusiasm. Regrettably, Tycho subscribed to the view of Ptolemy that the earth was the centre of the universe. Possibly to appease his many enemies at Court, he claimed to reject the then theologically heretical sun-centred theory of Copernicus (see below), and insisted that the sun with its train of planets circled around the earth. Fortunately, this rather weak compromise had no permanent influence on science.

Tycho carried astronomical observation as far as was possible with the quadrant only. His accuracy was remarkable, his readings being correct to about two minutes of arc. His tables of the motions of the planets and the sun were better than any that had gone before and he established the length of the year to less than a second. This showed the existing calendar to be causing a cumulative error and in 1582 Pope Gregory XIII agreed to a correction. Ten days were cut out of the calendar, producing an interesting and illuminating sidelight on human understanding. Thousands of people became convinced that their lives were being shortened and in their fury went on the rampage, rioting in the streets. It is easy for us to smile at this example of human folly, but these people were deadly serious and genuinely believed that ten days had been stolen from their lives. There are plenty of examples of similar irrationality, even today, among non-scientific people. Even the scientists are by no means entirely rational in their private lives, but any lapse of rationality in their work will be quickly detected and pointed out by their peers.

Other less major corrections to the calendar were also ordained and the Gregorian calendar is now in universal use. The respect with which Tycho's name was held by scientists was later shown when it was given to the most prominent crater on the moon. Ptolemy, and other astronomers, had to make do with smaller craters.

Ptolemy's book remained the sole scientific authority on astronomy until his beliefs were challenged by Copernicus, the real founder of modern astronomy. By careful observation he was able to compile accurate tables of the movement of the planets. These, together with his painstaking calculations, clearly indicated that, in defiance of Ptolemaic orthodoxy, the planets, including the earth, rotated around the sun. Copernicus incorporated his life's work into the book *De revolutionibus orbitum coelestium* (On the Revolutions of the Heavenly Spheres), completed in 1530 and published just before his death in 1543. A printed copy was put into his hands as he lay on his death-bed.

Copernicus's demonstration that the earth was a planet moving round the sun seemed to most people of the time, at best, foolish and perverse and, at worst, heretical and damnable. The earth was obviously stationary and the sun visibly moved round it, and, moreover, it said as much in the Scriptures. Where, in Copernicus's scheme, was Heaven? Ironically, Copernicus, who was an ecclesiastic, had dedicated the book to the Pope, and a Cardinal had actually paid the cost of printing it – presumably without reading it.

It took the Church some little time to realize that it had been nurturing a viper in its bosom. At the time, few could understand the learned explanations in this book. But when its central conclusions finally became apparent, the Church hurled anathemas at its author and at anyone who believed him. Had he lived, it is probable that he would have been burned at the stake, as were others who insisted that he was right. The Copernican cosmology was a revolution not only in science but also in human thought, and it was not until the eighteenth century that it was fully assimilated.

Modern cosmology is largely concerned with the problem of the origin of the universe. It is only in the twentieth century that cosmological theory has advanced sufficiently beyond the realm of vague speculation to provide anything approaching a plausible account. Today's cosmologists are no longer the fanciful dreamers of the past but are hard-headed scientists whose hypotheses are made on the basis of known facts. Their work has become very complex, involving astronomy, mathematics, physics, and, for some, philosophic speculation. At the same time it has become far more plausible, because it is based on an increasing volume of established fact.

The finding that triggered off modern cosmological thought was the discovery in the 1930s, by the American astronomer Edwin P. Hubble (1889–1953), that the light from all distant galaxies is coming from a receding source. This is the only acceptable explanation of the fact that the light coming from these galaxies is observed to be shifted towards the red end of the visible spectrum. The cause of this is the Doppler effect. You will be more familiar with this in the context of sound than that of light, but the principle is the same. When a moving sound source approaches you hear the pitch of the sound rising, and when the source passes and retreats the pitch falls. You can consider that the sound waves from the approaching source are 'compressed' so that more arrive in a given time, causing a raised pitch; and you can think of the waves from the retreating source as being 'stretched out'.

In 1948 the Russian-born American physicist George Gamow (1904–68), in a theory concerned with the origin of the light elements, interpreted this Doppler effect as demonstrating that the universe was expanding. Nearly everyone now accepts this as a fact. This expansion, however, implied that it had originated in a single point billions of years ago. This idea received wide support but was not widely known outside scientific circles until the British cosmologist and science fiction writer Fred Hoyle (1915–), in a highly critical and would-be destructive account, referred to it, satirically, as the 'Big Bang' hypothesis. Ironically, Hoyle's comment had quite the opposite effect to what he intended. Popular imagination and attention was caught by the phrase 'Big Bang' and soon everyone had heard of the theory.

Hoyle strenuously opposed the Big Bang idea. He pointed out that the original calculations of the age of the universe, based on Hubble's figures, indicated that it was younger than known geological evidence suggested. The theory also failed to account for the formation of elements heavier than helium. Hoyle and two other young Cambridge scientists, Hermann Bondi and Thomas Gold, had already, in 1946, proposed a completely different hypothesis for the origins of the universe – the steady state theory. This, they claimed, could account for the expansion without involving an unrealistic time-scale. The steady state theory proposed that the universe had had no beginning and that matter is continuously created at exactly the rate required to compensate for the expansion. In this way the overall density of the universe would remain constant. Critics suggested that this theory contravened the law of conservation of mass and energy. Hoyle replied that so did Big Bang.

The steady state theory became linked to the idea of nucleosynthesis – an explanation of how heavier elements might be made from hydrogen and helium. This theory soon commanded a great deal of support from many who found the Big Bang idea unpalatable. It received a severe setback, however, when observations at Mount Wilson observatory showed that the galaxies were much further away than had previously been supposed. This meant that the calculated age of the universe had to be increased to a minimum of 10 billion years – a figure entirely consistent with known geology and the Big Bang hypothesis. The steady state theory had another major snag – it predicted a zero temperature for space, which is absurd.

The final blow to the steady state theory came in May 1965. Two scientists working on radio astronomy at the Bell Laboratories, Arno A. Penzias (1933–) and Robert W. Wilson (1936–) were much troubled by microwave noise in their receiver. Initially attributed to pigeon droppings, every effort to discover the source of this interference failed. Eventually, careful tests showed that it was coming from everywhere in the universe. At this point it should be mentioned that radio waves, light, heat, X-rays, gamma rays and cosmic rays are all the same kind of radiation – electromagnetic. They differ only in wavelength. So there is nothing odd about picking up heat radiation with a radio receiver. All you need is one that can tune to the right very short wavelength.

The noise that Penzias and Wilson were detecting has a wavelength corresponding to the radiation emitted by a black body at a temperature of 2.74 kelvin (K); this is a very low temperature, just a little above the lowest possible temperature, absolute zero (0 K, or – 273.15°C). Penzias and Wilson went to see the physics professor Robert Dicke at Princeton University and were astonished to learn that Dicke had already predicted exactly such radiation. In fact, the people in his department were actually in the process of constructing a radio telescope to detect it.

Every body in the universe that is at a temperature above absolute zero emits electromagnetic radiation. These 2.74 K signals do not, however, come from cold bodies. They were actually emitted, a very long time ago, by white-hot matter moving rapidly outwards. On their way to us, travelling, like all electromagnetic radiation, at the speed of light, they have undergone so profound a red shift as to increase their wavelength to that corresponding to a microwave frequency. Calculations based on the parameters of the now standard model of the origins of the universe agree with numerous measurements made during and since the 1970s of the wavelength and uniformity of the 2.74 K radiation. When the news of the findings at the Bell Laboratories was published, support for the steady state theory collapsed. Today, no alternative theory to that of the Big Bang is seriously considered by the experts.

So now we have reached this stage of extraordinary sophistication in cosmology, can we congratulate ourselves that we have explained the origin of the universe? Certainly not. Explain all this to any bright child and he or she will say: 'Yes, but what was there before the Big Bang started?' Don't expect to get an answer to this question, even from the most egg-headed astrophysicist. There is no answer, and, for a very good reason, there is no point in trying to look for one in science. Science is concerned with demonstrable fact and deals only in matters of which we can have direct or indirect experience. Its function is to describe how things are. This gives it plenty of scope, but it also imposes strict limits. Science has nothing to say on matters that lie outside any possible human physical interaction because such matters do not provide us with the essential data that science uses.

To the scientist, when acting as a scientist, questions such as 'What was happening before the Big Bang?' actually have no meaning. You can ask them as often as you like, but you are just making pointless noises. As we have noted, scientists are, however, human beings and they share the human need for explanations. In contexts such as this, scientists, like anyone else, may fall back on non-scientific activity. They may experience religious faith, go in for spiritualism, propose elaborate theologies of their own, and so on. All this is fair enough, but scientists are the first to recognize that such things have nothing to do with science.

There is not the slightest indication that energy will ever be obtainable from the atom.

Albert Einstein (1879–1952)

They laughed at Wegener

Alfred Wegener (1880–1930) was a German weather expert whose claim to fame has nothing to do with weather. Wegener lived at a time when specialization in science was less firmly applied than it is today. In his day, especially in Germany, there was a tendency for the study of the earth (geology) and the study of its atmosphere (meteorology) to be lumped together. So it is not very surprising that Wegener took an interest in a subject that, today, would be considered completely foreign territory for a meteorologist.

Wegener actually started his scientific career by studying astronomy and his doctorate was awarded, in Berlin in 1905, for a thesis on astronomical work. He then took up meteorology and, between 1906 and 1908, he studied the climate of Greenland. He was then appointed to a lecturership in meteorology at Marburg. His book *The Thermodynamics of the Atmosphere*, published in 1911, became a standard textbook. In 1912 he went back to Greenland for a year but soon his work was interrupted by World War I, in which he saw active service.

Until nearly the beginning of the twentieth century everyone simply took for granted that the present position and general shape of the continents and of the oceans had remained unchanged from the time the earth's crust solidified thousands of millions of years ago. Obviously minor changes of coastline had occurred from sea erosion and, over the aeons, this could somewhat change the shape of a continent, but that was all. The idea that continents might be 'drifting' was first seriously put forward in 1912 by Wegener in his book *Die Entstehung der Kontinente und Ozeane*. This book was published in 1924 in an English translation entitled *Origin of Continents and Oceans*, and it met with outright scorn from the geological establishment.

The experts had little alternative but to laugh at Wegener and to rubbish his ideas; to do otherwise would be to throw aside some of their most cherished beliefs about the structure of the earth. The earth's crust was a continuous single structure, as solid under the ocean floors as elsewhere. It was inconceivable that continents could push their way along the ocean floors. So abuse was heaped on the unfortunate Wegener and heaped in abundance. Apart from the sheer fatuity of the idea of whole continents floating about, one of the pundits' main objections was that Wegener had not even tried to suggest any mechanism that could cause continents to move.

Fortunately, there were some open-minded people who found that Wegener's arguments had a good deal of cogency, and his suggestion was not altogether forgotten. What were these arguments? For a start, Wegener was impressed by the neat way in which, by cutting up a map of the world, you could fit the west coast of Africa into the east coast of South America. Furthermore, the north African west coast bulge fitted nicely into the hollow in the Caribbean and the east coast of North America. In fact, with a little imagination you could fit the whole west coast of the old world into the whole east coast of the new world. This, of course, could be pure coincidence or there might be another explanation that was a bit more plausible than floating continents.

But Wegener had other reasons. There was what appeared to be clear evidence that the continents actually were moving apart. If nineteenth-century longitude figures were to be believed, Greenland had moved a mile further away from Europe in the previous hundred years. So Wegener searched for further evidence of continental movement, and found that, on the evidence of existing figures, Paris and Washington were separating at a rate of about 5 metres (15 feet) every year.

Then there was the intriguing and unquestionable fact that identical fossil animals had been found on separate continents that were so far separated by oceans as to make it impossible that animals could have swum from one to the other. How was this to be explained other than on the basis of continental drift? If the animals couldn't move under their own steam from the present point A to the remotely situated point B, these points must once have been close together. It was a clear case of Mohammed and the mountain in reverse.

Wegener was a real scientist. Faced with facts of this kind, he felt obliged to propose a hypothesis, however improbable, that could account for them. So he proposed that continents were moving. He went further and suggested that, originally, there had been only one land mass – which he called 'Pangaea' (literally 'all–mother earth') surrounded by a single large ocean. Pangaea had then cracked in various places, and broken up into various segments which had slowly drifted apart. This, of course, allowed the ocean to flow between them to form new seas. Wegener suggested that these continental segments actually were floating on an underlying semi-liquid layer of molten volcanic rock and, over the course of many millions of years, had moved to the present familiar positions.

Having had a good laugh at what Wegener called a 'basalt ocean' his unkind critics now looked for something that would demolish his theory – and they found it. They were delighted to discover that Wegener's important data showing that continents were moving apart were actually based on other people's faulty determinations. Modern and more accurate methods had actually shown no evidence of movement.

This was a serious setback for Wegener but by now his partisans had also mobilized their forces and refused to give up. They pointed out that Wegener's idea explained much that was otherwise inexplicable and, in any case, did not depend on evidence of current or recent movement. The theory accounted, among other things, for changing patterns of glaciation and for the known change in the relationship of the poles to the land masses.

In 1924 Wegener was appointed professor of meterorology and geophysics as the first holder of a newly established chair at Graz University. But he never lost his interest in Greenland. Having planned a major expedition for 1930–31 he paid a third visit in 1929. In 1930 he was back for the fourth time. He was an intrepid man. In trying to cross from a camp on the central ice-cap to the expedition's base on the west coast, he ran out of luck altogether and lost his life.

Wegener died without ever knowing that his theory of continental drift was to be vindicated. The idea gradually gained hold and, as new and suggestive evidence, such as the structure of the continental shelves and the features of the mid-ocean rift, appeared, his theory became more and more widely accepted. In the early 1960s new evidence came to light from the study known as palaeomagnetism.

To explain this evidence, a slight detour is necessary. The earth's magnetic field is produced by the movement of liquid iron relative to the solid iron core. Other iron-containing rocks become magnetized, by induced magnetism, in the same direction as the earth's field, as they cool below the Curie point – the temperature below which magnetism is possible. When they solidify, their magnetism becomes fixed, thus providing a record of the orientation of the earth's magnetic North and South poles at that time. When rocks more than a few million years old are examined for their magnetic orientation, it is found that the positions of the earth's magnetic poles at the time they solidified were different from the present positions. The older the rocks, the greater the differences in position. These differences are much too great to be accounted for by the known shifts in the positions of the magnetic poles and the only plausible explanation is that the continents have moved relative to the poles. Fifteen–love Wegener.

Then there were the advances in knowledge of the earth's structure and how it acquired its present form – a discipline now known as plate tectonics. The term 'tectonics', used in its broadest sense, simply means 'building' or 'construction'. In this context it is used to refer to the processes by which the surface of the earth achieved its present particular structure. The adjective 'tectonic' refers to the distortions of the earth's surface caused by the forces operating under it.

One way of finding out what is under the surface of the earth is to see what happens to the vibrations caused by earthquakes. This was first done in 1909 by the Croatian geologist Andrija Mohorovicic (1857–1936). On studying the seismic charts of a Balkan earthquake Mohorovicic noticed that shock waves that penetrated deep into the earth arrived at the seismograph before those travelling near the surface. The only way he could explain this was to suppose that the outer crust of the earth must rest on a more rigid layer in which vibrations travelled faster. The distinction between the crust and this lower layer must also be sharp, and this separation, somewhere between 10 and 40 miles below sea level, is now known as the Mohorovicic ('Moho' for short) discontinuity. Under the deepest oceans, the Moho discontinuity is only three or four miles below the ocean bed.

Modern research, much of it using man-made shock waves from small explosions, has confirmed that the earth is actually layered like a giant onion – with an inner core, an outer core, a mantle and a crust. The mantle, in turn, is divided into the semi-solid mesosphere (Greek *mesos*, middle), the soft, semi-fluid and weak, asthenosphere (Greek, *asthenos*, weak), and the brittle, rocky lithosphere (Greek, *lithos*, stone). The mantle, the part of the earth between the crust and the core, accounts for more than 80 per cent of the total volume of the earth. Immediately under the earth's hard outer crust is the lithosphere. Under this lies the partially molten and weaker asthenosphere of the mantle. The plate tectonic theory proposes that the relatively brittle lithosphere, with its overlying crust, is broken up into slabs or plates as a result of the stresses caused by the convection heat currents from the hotter and more fluid underlying layer.

When the great oceanic ridges were discovered during the 1950s an explanation of continental separation became apparent. Scientists now believe that these ridges are the sites at which the hot rock (magma) is forced up through the relatively thin crust. The crust is, of course, thinner in the beds of the oceans. This magma cools and solidifies to create new oceanic lithosphere, but as it is forced upwards the solid lava must necessarily be moved sideways as more magma is forced upwards. In this way the asthenosphere is expanded sideways in this region, carrying the continents with it.

There is, of course, a finite quantity of magma in the asthenosphere and, as some rises to form new lithosphere, there must be a compensatory return of material from the lithosphere to the mantle. This occurs at what are known as subduction zones, most of which lie around the edges of the Pacific Ocean. At these zones expanding lithosphere is forced down into the mantle where it gradually melts. This is often accompanied by volcanic activity. Thirty–love Wegener.

By about 1960 the theory of plate tectonics had been established as one of the central principles of geophysics. Eight major plates and seven minor ones have been identified, and these plates can be though of as floating on the underlying semi-solid asthenosphere. Some plates are solely oceanic, but most include both ocean and dry land. None of them is exclusively continental. There are three types of boundary between plates. Constructive plate boundaries are spreading margins where new lithosphere is being formed. Destructive plate boundaries include those where lithosphere is moving down into the mantle, and those – called collision boundaries – where two plates push against each other and force up the crust to form mountain ranges and plateaux. Conservative plate boundaries, or transform faults, are those where plate edges slide past each other but are neither created nor destroyed. Fortunately, most of the conservative boundaries are on the ocean floor, because those that also involve land, such as the San Andreas fault, are a dangerous cause of crust movement (earthquakes). Forty–love Wegener.

As might be expected, volcanoes occur at plate boundaries. This is why volcanoes are often disposed in lines, as in the case of Hawaii (which was formed from five volcanos and the intervening volcanic ridges) and the Hawaiian-Emperor chain of islands and submarine volcanoes to the northwest. The line of the volcanoes is, of course, the line of the interface between two plates.

Game and set Wegener.

> The more important fundamental laws and facts of physical science have all been discovered and these are now so firmly established that the possibility of their ever being supplanted in consequence of new discoveries is remote.
>
> *Albert Michelson (1852–1931) the American physicist, whose experiment with Edward Morley (1838–1923) in 1887 changed the whole face of science and undermined the entire basis of classical physics*

The flat earth theory

We are sometimes apt to think that engagement in scientific thought is a recent activity. But this idea ignores the fact that humans are enquiring animals and are never really comfortable with unanswered questions. This has been so from the earliest appearance of human beings on this planet. As communication evolved, so did different kinds of grunts or gestures for why? how? what? who? when? and where? and we can be sure that these questions were constantly being asked. Moreover, once language was reasonably well evolved, the possibility of a crude kind of scientific thought and expression arose.

There are several ways of defining science, but, at this early stage, it was a matter of classifying objects, providing descriptions of them, and offering ever more satisfying explanations for observed phenomena. As we have seen, many of these explanations involved magic and crude religion. Primitive science involved many assumptions, which had to be gradually amended as more data accumulated. What is more natural than that simple people, living at the dawn of science, should assume that the earth was flat? Not completely flat, of course, because anyone could see that there were mountains and valleys, but generally flat. It may seem unkind to describe this quite natural idea as a blunder, but the fact is that from the beginning of human awareness, clear evidence to the contrary was available to almost everyone.

The earliest notions of the nature of the earth was that it was the floor of a kind of box, the sides of which were mountains that supported the sky. Round the outside of the floor flowed an enormous river on which was a boat that carried the sun. In some cosmologies, this box was at the centre of the first of a sequence of ever-larger concentric crystal spheres. The early Hindus decided that the earth was situated on the back of a large elephant, which was standing on the back of a giant tortoise, which was swimming in a sea of milk. The ancient Greeks held that the flat earth ended at the Pillars of Hercules and was surrounded by water occupied by a hundred beautiful young girls, the Nereids.

There were a few obvious objections to these theories and a number of indications that they wrong. The sun and the moon, for instance, were seen to be near-perfect circles and the only question was whether they were discs or spheres. If they were discs, why was it that they always presented the full diameter and were never seen edge on? It must have seemed fairly obvious that they were spherical, and, if this was so, why shouldn't the earth be spherical also? If the earth was flat, why was it that the horizon of the ocean showed a distinct curve? By the time boats with masts had been developed there was clear visual evidence that, as a boat approached from a distance, the first thing to be seen was the top of the mast, then the sails, and finally the hull.

If the earth was flat, it must have an edge. Why had no one ever reached the edge? Long after writing, and even printing, had been invented and records existed that dated back for thousands of years, there were still people who believed that the earth was flat. The fact that there was no record of anyone falling off the edge did not shake these people's confidence. 'No one would be so foolish as to approach too near the edge,' they said, 'There be dragons.'

> I can accept the theory of relativity as little as I can accept the existence of atoms.
>
> *Ernst Mach (1838–1916),*
> *Austrian physicist after whom Mach numbers are named.*

Getting the calendar wrong

The calendar is so familiar that we are rather apt to think of it as no more than a simple representation of the passage of time and a way of making dates for the future. But there is much more to the calendar than that, as a brief review of its history will show. In fact, <u>producing a good calendar is very much a matter of strict science</u> and, as such, has been the subject of a succession of blunders.

Once the need for a calendar became apparent, many centuries ago, for the regulation of social, business, agricultural and religious life, people had to decide how to organize it. The obvious start point was the day. No one can be in any doubt about days. If you measure from sunrise to sunrise, a day and a night become one unit. So the earliest calendars were simply a list of days numbered into the future. The fact that the sun didn't actually rise, but that the illusion was caused by the rotation of the earth, was at that stage irrelevant. Although the early calendar makers were unaware of it, they were actually assigning the term 'day' to a single rotation of the earth on its axis. So a day is the length of time it takes the earth to rotate once.

If the period of a day is thought to be too long for any particular practical purpose, you can divide it arbitrarily into as many subdivisions as you like and call them hours. These subdivisions don't even need to be all the same length, although it is probably most convenient if they are. We have become accustomed to 24 equal-length hours, which is quite a good arbitrary choice, as the number can be divided by 2, 3, 4, 6, 8 and 12.

To reckon into the future in terms of days gets to be tedious. It's fine for two or three days or even for the number you can count on one hand. But many more than that becomes awkward. So we need another arbitrary length of time, equal to a fixed number of days, so that we can conveniently represent longer periods of time. The seven-day week is also an arbitrary period of time. There are no natural events, like sunrise, to determine this time interval. The number seven, however, has long been considered rather special and is widely believed to have some kind of mystical or religious significance.

Our present weekday names are derived from early Anglo-Saxon and Norse mythology. Monday is Monandaeg, the day of the Moon; Tuesday is the day of Tiu, the son of Odin and a younger brother of Thor; Wednesday is Woden's, or Odin's, day; Thursday is Thor's day; Friday is Freya's day, Freya, or Frigg, being the wife of Odin and goddess of love; Saturday is Saeternesdaeg (Saturn's day); and Sunday is dedicated to the sun (Sunnandaeg).

With the month, the plot thickens and science becomes necessary. This is because the month is not an arbitrary unit of time for our calendar, but refers, of course, to the full cycle of the phases of the moon – new, full, crescent, half, gibbous (more than half). When it is waning the horns of the crescent are towards the west; when it is waxing they point to the east. And this is where the trouble starts. The five phases of the moon take 29.53059 days to complete. This is called the synodic month and it is, obviously, far from being an ideal timer for the calendar month.

So far as the next division of time – the year – is concerned, there is another problem. The ancients quickly recognized that you could, very nearly, fit 12 lunar months into one complete cycle of the seasons. But, it turns out that there are several different ways of measuring the year. So, for a start, you have to decide what you mean by a year. One good way is to reckon the year as the time between one vernal equinox and the next. The vernal equinox is the day in the spring when the night and day are of equal length. There is also an autumnal equinox.

If the year is decided in this way, it is found to contain 365.242199 days, or, in other words, 365 days and nearly six hours. Thereby hangs a major snag for people trying to put a calendar together. Twelve lunar months add up to 354.35706 days, i.e. nearly 11 days less than the year between the equinoxes. So if you base a calendar on lunar months, you will find that there is no permanent relationship between the month names and the seasons. A month, like August, that starts out in the middle of the summer will, in the course of about 16 years, end up in the middle of the winter.

The obvious solution to this one is to change the length of various months from 29½ days to 30 or 31 days. This still doesn't get it quite right, so we have another fine adjustment every fourth February – the familiar leap year, when that month gets an extra day.

Bearing all this in mind, it is hardly surprising that the early calendar compilers had a hard time. About 600 BC the Romans had a 10-month calendar that turned out to be hopeless. So they produced a 12-month one based on the lunar month, giving a year more than 10 days shorter than our present year, which is based on the equinoxes. Their idea of how to keep this year in step with the seasons was to slip in an extra month, every second year, between 23 and 24 February. Unfortunately, the authorities couldn't get their act together on this one and, half the time, they forgot to announce the extra month.

The result was that by the time of Julius Caesar (100–44 BC) the calendar was in a hopeless mess. So Caesar appointed an expert called Sosigenes to look at the problem. Fortunately Sosigenes was an astronomer and had some idea of what was going on. He suggested that they should drop the lunar month basis and replace the year with the equinox system of 365¼ days. This was a great advance, but there was still the difficulty of the accumulated errors of the previous 500 years. To try to sort this out, Sosigenes suggested that the year 46 BC should be lengthened by 90 days. This put 1 January somewhere in what was, at the time, the middle of March. Sosigenes also proposed adding the additional leap year day in February.

The Senate were pleased with these proposals and agreed them. It was a popular move with the people also, and the senators hardly thought it necessary to point out to them that their lives were not actually being lengthened. Julius only just lived to see the new system inaugurated before meeting the fate on the Ides of March so eloquently described by Shakespeare. However, he still got the credit for the calendar, and for 1600 years the Julian calendar was the standard for most of Europe. There was still a snag, however. The Julian calendar year was 365.25 days long, while the equinoctial year was ever so slightly shorter at 365.242 days. Every year that passed means an additional error of just over 11 minutes.

So, by 1582, there was a discrepancy of 10 days. Pope Gregory XIII now got into the act and published a papal bull that had been drafted by a Jesuit astronomer called Christopher Clavius. The effect of this was that in October of that year 10 days were cut out of the calendar and <u>everyone was suddenly 10 days older. As we have seen, this caused riots.</u> It seemed that nearly everyone actually believed that their lives were being shortened by 10 days and they bitterly resented it. It was some time before the furore settled.

Clavius redefined the length of the year as 365.2422 days. This new Gregorian calendar (note that the Pope gets the credit, as did Julius Caesar) was much better than the old Julian calendar. It produced an error of just over three days every 400 years. But Clavius had an answer to this problem also. He suggested that although the last year of each century would normally be a leap year (being multiples of 100, which is divisible by four) three out of every four should be considered a non-leap year. So 1700, 1800, and 1900 were non-leap years. The year 2000 will be a leap year.

This was an extraordinary feat of calendar-making. The Gregorian calendar, which is in use to this day, is remarkably accurate; the <u>difference between the Gregorian year and the solar year is less than half a minute.</u>

To us who think in terms of practical use, the splitting of the atom means nothing.

Lord Richie Calder (1906–82), writing in 1932

Why is the sky dark at night?

Here's a paradox for you. It isn't new; in fact it has worried people for hundreds of years. If the universe is infinite and contains an infinite number of stars, then the whole of the night sky should be brighter than day. In whatever direction you look, your line of sight should end in a star. Although the brightness from each star decreases in proportion to the square of its distance from the earth, the number of stars must increase in proportion to the square of the distance from us. Consequently, the whole sky should be as bright as the sun.

This seemingly extraordinary contradiction is known as Olbers' paradox, after the German astronomer Heinrich Olbers (1758–1840), who in 1826 posed the seemingly naive question: 'Why is the sky dark at night?'. Olbers gets the credit, but, as usual, many other people previously had wondered about this question, including J. P. L. Chesaux in 1744, Isaac Newton and his friend Edmund Halley (of comet fame), and Johannes Kepler in 1610. None of them had any answer to it.

The resolution of the paradox relies on identifying some underlying assumptions of the paradox that are wrong. The paradox is resolved in the Big Bang cosmology, in which the universe is not infinite; neither is it uniform or static. Two effects then come into play: the age and expansion of the universe. The finite age of the universe limits the light emitted by stars – there has been insufficient time since the Big Bang to fill the universe with starlight. Even in an older universe, the sky would remain dark because the lifetime of luminous stars is too short. There is also a small effect due to the expansion of the universe: light from receding galaxies is redshifted into the infrared part of the spectrum invisible to the naked eye (see Star errors).

> 1858 has not, indeed, been marked by any of those discoveries which at once revolutionize, so to speak, the department of science in which they occur.
> *Thomas Bell (1792–1880), President of the Linnean Society, speaking of the year in which Darwin read his papers on the origin of species to the society*

Lord Kelvin's time-bomb fizzles out

Among the many severe critics of Darwin was one of the outstanding physicists and mathematicians of the day Sir William Thomson (1824–1907), later ennobled as Lord Kelvin. Thomson went to Cambridge at 16 and was a professor of mathematics by the age of 22. His proposed absolute temperature scale, now named after him, is in universal scientific use. He worked out the second law of thermodynamics – that heat always passes from a point of higher to a point of lower temperature unless work is done to prevent it. He made a fortune from his invention of the highly sensitive mirror galvanometer, and he was a prime mover in the laying down of the first trans-Atlantic submarine cable.

Thomson's status can be judged from the fact that when he died he was buried in Westminster Abbey alongside Isaac Newton. Whether or not his opposition to Darwin was initially based on strictly scientific grounds is hard to say, but he certainly went out of his way to find reasons to deny Darwin's theory of evolution. One of the principal objections he put forward – and this was one that gave Darwin enormous pain and distress – was that there had not been enough time since the crust of the earth hardened and cooled for the processes of evolution to occur.

At the 1861 meeting of the British Association for the Advancement of Science, Thomson stated that it was highly improbable that the sun had illuminated the earth for as long as 100 million years. He thought that the most likely figure was 98 million years. This conclusion he also recorded in his *Textbook of Natural Philosophy*. Natural philosophy is what we now call physics. Thomson's case against Darwin rested on the reasonable assumption that all the heat of the earth, from its origins to the present, came from the sun. So, if one could calculate how long the sun could possibly have been burning, one had a measure of geological time.

Thomson developed a formula that took into account the mass and radius of the sun and the gravitational constant. His figure for the duration of the sun's burning suggested that it only had a few million years of life left in it. Darwin was shocked, because Thomson's figure for the age of the earth eliminated the possibility that his theory of evolution could be correct. There simply would not have been time for the geological processes on which Darwin's theory was based to have occurred. If Thomson was right, the earth must have been created with a ready-made geology. Archbishop Ussher's date of 4004 BC could be right.

Darwin was no mathematician, and here he was up against probably the most distinguished mathematician of the day. In desperation, he turned to his son George who was a mathematician. George laboured long on the problem and, having checked and rechecked Thomson's figures, was forced to conclude that the great mathematician was right. Collapse of stout party.

Unfortunately for Darwin, neither he nor Thomson was aware of certain rather important scientific facts that came to light at the turn of the century. In essence, this was that the core of the earth contained enough heat-producing radioactive material to totally upset Thomson's calculations and put the age of the earth a very much longer way back than anyone had ever suspected – more than enough to accommodate the whole of known geology and evolution to date. Radioactivity was discovered by the French physicist Antoine Becquerel (1852–1908) in 1896. As for the age of the sun, later scientific discoveries were to show that its continuing energy output was the result, not of ordinary chemical burning as if it were made of coal or wood, but of nuclear fusion. In effect, the sun is a kind of ongoing controlled hydrogen bomb.

Neither of these facts could, of course, have been known to Professor Thomson. However, aware that there were some highly dubious elements in his equation, he ought to have kept an open mind. Furthermore, there were other very good reasons, which *were* available to him, to accept Darwin's theory had he kept a properly open mind. All of which indicates that scientists, however distinguished, are still human beings prone to take up positions determined by prejudice and other non-scientific factors. In this case it was a simple religious piety that took the Bible's creation story as literal fact. It has to be said, however, that Kelvin was a man of supremely good nature, kind-hearted and modest, who, although holding to his opinions with unrelaxing tenacity, did so with unvarying courtesy. Thomas Henry Huxley's opinion of him was: 'Gentler knight there never broke a lance.'

Happily, George Darwin lived long enough to become aware of the reasons for his earlier inability to fault Thomson. In September 1903, he wrote a letter to *Nature* in which he stated: '. . . we have recently learnt the existence of another source of energy, and that the amount of energy available is so great as to render it impossible to say how long the sun's heat has already existed, or how long it will last in the future. Prof. Rutherford has recently shown that a gram of radium is capable of giving forth 10^9 calories. This energy is nearly forty times as much as the gravitational lost energy of the homogeneous sun . . .'

George, who had meantime become Sir George, and was professor of astronomy at Cambridge, was assuming that the sun was made of radium – a totally wrong assumption, but one rather nearer the truth than Thomson's assumption that it was made of coal. It might be hoped that Lord Kelvin would have had the grace to apologize, but, regrettably, although the evidence for radioactivity was plain, he refused to accept it. To do so would have been to give up his opinion as to the age of the earth and perhaps, even, be forced to agree with Darwin on the subject of evolution.

All this talk about space travel is utter bilge, really.
Richard Woolley, British Astronomer Royal, 1956 (five years before Yuri Gagarin made the first space trip)

The CIA and UFOs

In August 1997, the American Central Intelligence Agency came clean and admitted that they had lied about UFOs in the 1950s and 1960s. Many of the innumerable public claims of sightings of unidentified flying objects were, after all, unidentified flying objects, not hallucinations or misinterpretations of natural phenomena. They were, in fact, high-altitude secret spy planes. It was not clear why, at this particular juncture, the CIA should have decided that confessional valour was the better part of discretion. Organizations like the CIA thrive on secrecy and are, presumably, well accustomed to a long-term tight-lipped position. So why come out with a revelation that is at least as embarrassing as others they have sought to conceal but could not?

There were several possibilities. One was that the CIA authorities were aware that the truth was about to break and decided to make a virtue of necessity and at least get the credit for seeming to be candid. Another possibility was that they had been infected with the confession bug, currently endemic in the States, and were pushed over the edge by the example of President Clinton's apology over the astonishing Tuskegee, Alabama, affair (*see* The syphilis scandal).

Alternatively, with the ending of the cold war, the CIA establishment might have been feeling a little insecure, perhaps even unwanted, and may have felt the need to display some human qualities. Perhaps more plausible is the idea that the confession was the result of the operation of the invaluable 30-year rule – a rule that ensures that no one currently admitting former sins could possibly have been responsible for the policy that led to them. Anyone who was in a position of any importance in the organization 30 years before was, by now, either demented or dead.

One would like to think that people in authority in the United States are becoming seriously worried about the growing epidemic of pseudoscientific insanity manifested by belief in UFOs, little green men, the acupuncture meridians, homoeopathy and the imminent end of the world brought about by malicious computer programmers. If this is the reason behind the CIA's uncharacteristic recent conduct, one can but applaud.

But there is a sinister alternative. Could it be that the CIA was being really Machiavellian and was actually trying to *revive* public interest in the UFO phenomenon? Clearly they had found this particular public fantasy extremely useful in prior decades. Perhaps they were anxious that it should not be forgotten. Well aware that no one outside the organization ever believes a word they say, what could be more stimulating to the UFO freaks than to produce a plausible-seeming explanation for the sightings? Since this explanation must *ipso facto* be a lie, great weight would be given to the proposition that UFOs must actually exist. *Quod erat demonstrandum.*

Evolution

I am strongly opposed to Charles going on this *Beagle* voyage. He is moving away from the Church, drifting irretrievably into a life of sport and idleness.

Robert Waring Darwin (1766–1848),
Charles Darwin's father

The great Lamarck blunder

Jean Baptiste Pierre Antoine de Monet, Chevalier de Lamarck (1744–1829) has, quite unjustly, been the butt of biology students' jokes for years. This is because he is remembered, not for his numerous contributions to natural history, but for a monumental error that causes smiles of patronizing superiority in those who are being wise after the event.

Lamarck, although of aristocratic parentage, was born poor and was destined for the church. He was attending a Jesuit college when his father died and, at the age of 16, he dropped out and joined the army. An injury caused him to leave the army and he took up the study of medicine. He soon found botany more interesting, however, and wrote a best-selling book on the identification of plants. This led to his being appointed Botanist to the King of France at the age of 37. Later he turned to zoology. Encouraged by the success of his botany book, which went through several editions, he now produced a book *Zoological Philosophy*, published in 1809, and then a major work on invertebrate animals called *Histoire naturelle des animaux sans vertèbre*. The first volume of this was published in 1815.

At this date in the history of science many well-informed scientists already accepted the general idea of evolution. Most people outside science, however, still believed that all species were produced, fully developed, by a single act of creation. This was what the Bible said and to doubt it was heresy. A few had other ideas. Some believed that species appeared by a process of 'spontaneous generation' especially in rotting meat and vegetation. Others attributed the creation of new species to God. Ideas became more regularized when scientists like the Swedish naturalist Carl Linnaeus (1707–78) began to publish books in which different species were arranged and classified and it became apparent that there were close similarities between certain species. Those of enquiring mind began to wonder whether there might be some kind of relationship between species that was more than mere resemblance.

There was another major source of speculation about this. For many decades geological scientists had been busy determining the structure of the earth. In the course of this work, two important facts emerged: that stratified layers of rock had been formed at different periods in the past, and that the fossils found in these different layers were the impressions of creatures that lived during these periods. From these observations the science of palaeontology – the study of prehistoric life based on the evidence of fossils – rapidly developed.

People had, of course, known about fossils for thousands of years and ideas about them varied. The Greek philosopher Aristotle taught that they were failed or abortive attempts at spontaneous generation from mud. Leonardo da Vinci and other Renaissance thinkers, however, recognized them for what they were – the impressions left by once-living organisms. Although they discussed this probability, they had to be careful as such ideas were seen as heretical by the Church.

The most impressive finding of the palaeontologists was the record of definite sequences of animals, living at successive periods but showing general resemblances to their predecessors. The older the rocks in which they were found, the simpler and more primitive were the life forms. This strongly suggested that modern living things came from earlier and less complex life forms by a gradual process of change – by evolution. Although this seems obvious to us now, the influence of formal religion was still very strong at that time and many people were deeply disturbed by evidence that seemed to contradict what it said in Genesis, the first book of the Bible.

Science has shown that the earliest known fossils, similar to blue-green algae, date back about 3000 million years. Sediments laid down about 1500 million years ago contain large numbers of single-celled organisms (protozoa). Sediments dating back about 590 million years contain the imprints of soft-bodied animals such as segmented worms, and in deposits made about 570 million years ago you can find fossils of molluscs, jellyfishes, crustaceans and starfishes similar to those existing today. Fossils of many species of fish and of jointed creatures called trilobites and large scorpion-like animals are found in deposits dating back 400–500 million years. Fossils of amphibians and insects appear at about the same period.

The fossil record tells us that the first reptiles appeared about 340 million years ago, giving rise to the dinosaurs, which dominated life on earth from 230 to 65 million years ago. The earliest mammals emerged at about the same time as the dinosaurs, while the birds are thought to have evolved from one group of dinosaurs around 200 million years ago. Fossils of the first primates – the group of mammals that contains the living lemurs, monkeys, apes and humans – are found only in deposits dating from about 65 million years ago. Remains of recognizably human-like creatures occur only in deposits laid down in the past few million years.

All this evidence was available in the early decades of the nineteenth century and it must have caused many people who knew about it to think furiously. But there was also strong evidence from another scientific discipline, the evidence of body structure. As knowledge of the structure of the human body and of those of other species grew, it became apparent that the structural resemblances were much closer than had been thought. When the anatomical structure of animals with backbones was studied, it was seen that animals differed essentially only in the shape and size of individual parts rather than in possessing totally different parts. The similarities between even quite remote vertebrate species were far greater than the differences.

Equally striking was the finding that many animals possess vestigial structures that no longer serve any purpose but that were present, in functional form, in other species. Ostriches have rudimentary wings but cannot fly. Snakes have vestiges of hind-limb bones. Whales have a vestigial pelvic girdle. Humans have an appendix attached to the large intestine which corresponds to a blind-ended digestive structure in herbivorous animals, and a vestigial tail in the form of the coccyx. Some marsupial mammals, which have been delivering live offspring for millions of years, still show embryonic indications of the egg tooth with which the young of various egg-laying species crack the egg-shell. All these observations strongly suggested common ancestry and some kind of evolutionary process.

With this background, we can now return to Lamarck. He was

aware of all this evidence and was determined to find an explanation for all these suggestive facts. So in his 1809 book he put forward the idea that animals and plants produced changes or even new organs in response to a need resulting from changes in their environment. On this basis he might suggest that giraffes, for instance, acquired long necks because of the need to stretch up to reach the leaves at the top of trees. If mice suffered a disadvantage by having their tails caught in something, their tails might grow shorter, or even disappear altogether. The implication of this was that tail-less mice could be produced by repeatedly cutting off the tails of mice and then letting them breed. This idea was further developed in Lamarck's natural history book of 1815.

Lamarck's ideas of evolution were not widely accepted and, indeed, very little attention was paid to them until after Darwin came up with his theory of evolution. It was then that everyone began to laugh at the unfortunate Lamarck. In fairness to him, it must be said that some of the criticisms levelled at him were grossly unfair. His use of the word *besoin* (need), for instance, was mistranslated to suggest that he meant that animals had the *desire* for certain changes. There was no question of this.

Lamarck's misfortune was that Darwin's theory fitted the observed facts so well. Everyone who understood it, and who also knew what Lamarck had taught, saw at once that Lamarck must be wrong. The principal difficulty for Lamarck's theory was to explain how the changes experienced by plants and animals could be passed on to the next generation. Lamarck was well aware that physical characteristics were transmitted by eggs and sperm so this was a real difficulty. His answer to this was the rather feeble suggestion that bodily changes could somehow modify the sperm or the ova in such a way that the new characteristics were passed on to the offspring.

We now know that the genetic information in eggs and sperm is acquired by the new individual at the moment of conception. It is, in fact, simply the result of the combination of the DNA in the father's sperm and the DNA in the mother's egg. The resulting genome is present in every cell of the new individual's body including the cells in the testes and ovaries from which the new individual's sperm or eggs are formed. So there is no way in which gross bodily changes acquired during the lifetime of the individual can so affect its DNA that its offspring inherit the changes.

A second problem with Lamarck's theory was that observable fact did not support his theory. Cutting off the tails of mice for scores of generations will never breed a strain of mice without tails. Acquired characteristics are not passed on.

It must not be thought that Lamarck was a fool. He was, in fact, a most important figure in the history of biological science. It was Lamarck who first made the essential distinction between vertebrates and invertebrates; his classification separated spiders and other creatures from insects, and he recognized that these were distinct from various small crustaceans. He was also convinced that species changed because of environmental factors. In addition to all this, he coined the term 'biology'.

This story is not about Darwin, but the full flavour of Lamarck's blunder will not be appreciated unless you are clear about Darwin's idea. If you are clear, you can move on to the next chapter right now. Although Darwin gets all the credit, the current theory of evolution was put forward almost simultaneously by two men, Charles Darwin (1809–82) and Alfred Russel Wallace (1823–1913). Darwin and Wallace first outlined the theory in a joint communication to the Linnean Society in London in July 1858 but the real impetus to the spread of the new ideas came with the publication in November 1859 of Darwin's book *On the Origin of Species by Means of Natural Selection, or the Preservation of Favoured Races in the Struggle for Life.*

This book made Darwin famous, and Wallace's contribution, which ranked with Darwin's, was largely ignored.

Darwin got the first inklings of his theory while on a voyage of scientific exploration in HMS *Beagle*, many years earlier. On the Galápagos Islands Darwin observed 14 different species of finches, differing mainly in the shape and size of their beaks, none of which was known to exist anywhere else in the world. Each species occupied its own island. Some of the finches were seed-eating, others lived on insects, and their beaks differed appropriately. These observations set Darwin thinking. It seemed highly probable to him that a single species of finch – a seed-eating bird common on the nearby mainland – must have colonized all 14 islands a very long time before and that their isolated descendants had evolved into different forms in environments in which they were not in competition with other birds.

This idea was left to germinate in his mind. Later, in 1838, Darwin read a book by Thomas Robert Malthus (1766–1834) called *An Essay on the Principle of Population.* In this book Malthus suggested that human populations always grow faster than the available food supply and that environmental factors such as starvation, disease or war are necessary to limit them. Darwin was reminded of the finches. It occurred to him that the first birds to occupy the islands would have multiplied unchecked until there were too many of them for the available seed supply. Many would have died, but a few of them, who happened to be able to adapt to a different diet such as insects – perhaps by an accidentally and naturally occurring variation such as a differently shaped beak – would flourish and multiply, until they in turn outstripped their food supply. Some purely chance variations might provide good adaptation to the current environment and thus enable their possessors to survive to breed and pass on the favourable characteristics to their offspring. Other variations would prove unsuitable and their possessors would die before they could breed.

Those organisms that were naturally most fit survived and flourished. Over the course of millions of years this process, operating very slowly, could be seen inevitably and automatically to give rise to radical alterations in the characteristics of organisms. The different types that evolved in this way had shared common ancestors in the past.

New species could evolve by the splitting of one species into two or more different species, largely as a result of geographical isolation of populations. Such isolated populations would experience different environmental pressures and would evolve along different lines. If isolation prevented interbreeding with other stock derived from the same ancestors, these differences might become great enough to establish a new species that could not successfully breed except with its own kind. A species is, by definition, incapable of breeding successfully with another species.

This was the great idea of how changes in species occurred by 'natural selection', and Darwin saw that this principle alone was sufficient to explain evolution. Darwin never understood how heritable changes occurred or how the necessary spontaneous variations in species (mutations) occurred, but this did not in the least detract from the power or persuasiveness of his theory. To most unbiased scientists of the time, Darwin's idea, once fully understood, seemed self-evident.

Darwin was never able to explain why breeding could not occur between members of different species. Today, of course, we know that different species have incompatible genes and often even different numbers of chromosomes. As a result, the accurate matching up of chromosomes from the father with those from the mother cannot occur. Cells require matched pairs of chromosomes to function normally.

Darwin also had the problem that the fossil record of evolution does not show, as might be expected, a steady, gradual process of change from one species to another. Instead, change appears to occur in sudden jumps. New species appear to enter the picture suddenly and to change little during their term of existence. In the Precambrian era, ending about 570 million years ago, only algae-like creatures and the simplest invertebrates are found. Suddenly, at the start of the succeeding Cambrian period, there is a profusion of invertebrate groups. Darwin believed this to be a false impression arising from insufficient data, but we now have a better explanation.

Initially the earth's atmosphere had almost no oxygen, and it was not until plants became abundant that oxygen was able to accumulate. Photosynthesis – the natural production of organic compounds from carbon dioxide and water using light energy, with the release of oxygen – originated in single-celled organisms, and these evolved into plants. It may be that this fundamental change occurred around the late Precambrian era, thereby allowing the explosive proliferation of oxygen-breathing animals.

As for the source of spontaneous variation needed for natural selection to work, we now understand, in considerable detail, how this arises. Differences in genes and even in chromosomes (mutations) can occur spontaneously or be induced by radiation and other agents. In addition to mutations, egg and sperm production involve a complex reshuffling of parts of the chromosomes and even the loss or gain of whole chromosomes. Variation within a species is, of course, commonplace. Consider humans and how widespread are the differences in skin and hair colour, height, weight, body type, intelligence, and so on. Although we now have enormous power over our environment and can interfere with many of the natural selective processes, these processes are occurring all the same. The lifetime of any individual, or even of many generations of people together, are insignificantly short periods in the context of evolution, and changes are not apparent. This makes it difficult for us to perceive that we are subject to the same evolutionary forces we readily accept as acting on other species.

Every species, human included, produces over the course of very long periods, a large number of mutant types. If a mutation occurs in egg or sperm DNA, and, if the organism or person survives, this mutation is passed on through the generations. Some of these changes are so damaging that the affected organisms do not survive long enough to breed and the mutation is lost. Many are of little value in assisting in the process of adaptation to changing environments. A few, however, so change the individual that the power of survival is enhanced. If the characteristic caused by the mutation offers an advantage to the individual, a single such individual can, in the course of a number of generations, give rise to a very large number of individuals possessing the desirable new characteristic.

Major changes of this kind are very rare, and it is not surprising that we have recognized none in ourselves in the course of recorded history. The span of recorded human history – a mere 10,000 years or so – is just a blink of the eye in comparison with the whole period of the evolution of living things stretching back hundreds of millions of years. Although we can to some extent insulate ourselves from the effects of natural selection, and modify its operation by improvements in hygiene, nutrition, medical treatment and contraception, there is no reason to believe that we are immune to evolutionary processes.

The human generation averages about 20 years and evolution requires thousands of generations. This is so with all species, but there are some species with such a short generation that if we study populations of these we can actually see evolution in action. Such a group are the bacteria, some of whom have a generation of only 20 minutes. Bacteria can thus experience more than half a million generations in the time taken for one human generation. If bacteria find themselves in an environment containing an antibiotic, most of them will be killed. But there will be a few which, by pure chance variation or mutation, are able, to some extent, to resist the effect of the antibiotic. These will survive especially if the antibiotic is present in less than the proper dose, or for less than the proper period of time. Now, if all the susceptible bacteria are killed and only those inherently able to live with it actually survive, the latter will breed and will produce a new strain of organisms. This strain will be antibiotic-resistant.

This is no theoretical picture; it is happening all the time. Bacteria do become resistant to antibiotics and this is the mechanism – natural selection – by which it occurs. These evolutionary changes in response to environmental changes are seriously worrying to doctors and pharmacologists, who must constantly be developing new antibiotics if they are to be able to go on successfully treating severe infections in their patients.

Darwin would have sympathized but would have been secretly pleased to see the direct proof of his theory. It is hard to imagine what Lamarck would have thought.

> I doubt whether Darwin will have the determination to survive a difficult journey of several years. My studies of physiognomy indicate that people with a broad, squat nose like his don't have the character.
>
> *Robert Fitzroy (1805–65), Captain of the Beagle and Governor of New Zealand*

Soapy Sam's defeat

Charles Darwin was a shy, retiring man who was reluctant to confront his opponents. Fortunately, he had, as an ally and supporter, one of the most energetic and pugnacious scientists of the day, the redoubtable Thomas Henry Huxley (1825–95). Huxley was a tough and high-principled scientist but warm and friendly. There is a charming story of the occasion when his two grandchildren, Julian (later a distinguished biologist) and Aldous (later a notable novelist) brought him an insect they had concocted by glueing together various parts from other dead insects, and asked him to identify it. The great naturalist examined the creature with meticulous care, looked searchingly at the small boys and said solemnly: 'Tell me. When you caught it, did it hum?' 'Why yes.' said Aldous. 'Then I'll tell you what it is,' said Huxley, 'It's a humbug.'

When Darwin's *Origin of Species* appeared in November 1859, Huxley was delighted. Darwin's claim that his book would throw light on the origin of humans and their history 'was not only in full harmony,' he said, 'with the conclusions at which I had arrived respecting the structural relations of apes and men, but was strongly supported by them.' Huxley decided that as Darwin was not a specialist in development and vertebrate anatomy it would not be intruding on his province to deliver some popular lectures on the subject. 'In fact,' he said, 'I thought that I might probably serve the cause of evolution by doing so.'

Huxley's six lectures at a working men's club helped to bring the matter to the fore, and by the time the 1860 meeting of the British Association for the Advancement of Science was convened at Oxford, this was the hottest topic of the day. The question of the descent of man from the apes – a popular misunderstanding of the true theory – caused outrage to many and especially to the religious hierarchy who saw it as a blasphemous denial of the truth of the Bible. What Darwin was saying was that both the other apes and humans had a common ancestor from which both had descended. Today we recognize that humans have very many physical characteristics in common with apes, and humans are now officially classified as apes.

One member of the upper echelons of the Church took it on himself to be their spokesman on the matter. This was the eloquent and articulate Samuel Wilberforce (1805–73), Bishop of Oxford, whose social skills and popularity as a speaker had earned him the nickname of 'Soapy Sam'. Wilberforce had taken First Class honours in mathematics at Oxford in 1826 and, although he had had a distinguished career as a churchman, he was popularly regarded as both qualified and entitled to put Darwin and Huxley in their places. So Wilberforce not only wrote an article on the subject in *The Quarterly Review*, but let it be known that he would attend the meeting and that he intended to provoke a debate that would 'smash Darwin'.

The 35-year-old Huxley was at the British Association meeting but had no particular appetite for a debate on the evolution question. It seemed to him probable that with a mixed and largely non-scientific audience, to whom speakers would appeal on the grounds of prejudice rather than science, and who might have difficulty in grasping the real grounds of the scientific case, Darwin's position would hardly get a fair representation. So Huxley, who was also tired after a busy week, resolved to leave Oxford on the Saturday morning – the day of the debate. The previous afternoon, however, he chanced to meet another Darwin supporter called Robert Chambers who, on hearing of Huxley's intention, begged him to stay. The ever-obliging Huxley agreed to stay and attend the meeting.

When news went around that Wilberforce was about to speak, everyone tried to get into the lecture theatre. There was such a crowd that the meeting was moved to the long west room of the University Museum. Some 700 people managed to get in; even the window ledges were occupied. The platform was at the east side of the room between the two doors. On it, in the centre, was the President of the section, Professor Henslow. On his right hand sat Bishop Wilberforce and others of the ecclesiastical group, and on his left were Huxley, Joseph Hooker, Professor Beale of King's College, London, and various other scientific big names. Numerous clergymen, of course, were to be seen in the audience and many undergraduates had been attracted to the meeting by the hope of fireworks.

Before the discussion on evolution began, the chairman announced, very sensibly, that only those with valid points to raise would be allowed to address the meeting. This turned out to be a wise precaution as no fewer than four consecutive participants had to be told to be quiet because they were clearly interested only in upholding the religious viewprint and disparaging the other side. Eventually the Bishop stood up and the room fell silent.

It was at once evident to Huxley that Soapy Sam had been well briefed by a scientist – it was later admitted that this was Sir Richard Owen (1804–92), a real pillar of the old-guard scientific establishment and a long-term enemy of Huxley's. Owen was implacably opposed to Darwin's theory and bitterly resented Darwin's fame. More importantly, it was soon apparent to Huxley that Wilberforce had failed to grasp the central point of Darwin's theory (see The great Lamarck blunder). 'Soapy' was a most eloquent and entertaining speaker and easily carried the meeting along with him, but Huxley saw at once that the Bishop did not know how to manage his own case.

For half an hour Wilberforce spoke with supreme mastery of the language. With studied courtesy he ridiculed Darwin and savaged Huxley. Buoyed up by the enthusiasm of the audience and, as is often the case with demagogues, possibly pushed by exuberance into indiscretion, he came to what he no doubt took to be a brilliant conclusion by turning to Huxley and saying to him: 'Tell me,

Professor Huxley, is it on your grandfather's side or your grandmother's side that you are descended from a monkey?'

Huxley turned to his friend Sir Benjamin Brodie who was sitting next to him, slapped him on the knee, and said to him in an undertone, 'The Lord hath delivered him into mine hands.' Brodie was astonished and had no idea what Huxley was getting at, but, as Huxley wrote later: 'The Bishop's remark had justified the severest retort, and I made up my mind to let him have it.'

As the Bishop sat down, Huxley rose slowly to his feet, waited for silence and carefully explained Darwin's theory. Then, according to an account recorded by his son, Leonard Huxley, he said quietly: 'A man has no reason to be ashamed of having an ape for his grandfather. But if there were an ancestor of whom I should feel shame it would rather be a *man* – a man of restless and versatile intellect, endowed with great ability and a splendid position – who, not content with success in his own sphere of activity, plunges into scientific questions with which he has no real acquaintance, only to obscure the truth by rhetoric, eloquent digressions and skilled appeals to religious prejudice.'

The effect was tremendous. The audience shouted and applauded. One lady fainted and had to be carried out. When the commotion had settled, a call arose for a speech by Huxley's close friend Joseph Hooker (1817–1911). The President agreed and Hooker calmly and clearly demonstrated that the Bishop, by his own showing, had never grasped the principles of Darwinian evolution. The Bishop wisely remained silent and the meeting broke up. The Bishop's party filed out poker-faced, some doubtless thinking that their spokesman had justifiably been hoist with his own petard.

Huxley was the hero of the hour. Members of the audience crowded round and everyone wanted to congratulate him. Even several of the clergy joined in to express their approval. That evening one person said it had been so good he would like to have it all over again. Huxley 'with the look on his face of the victor who feels the cost of victory' said: 'Once in a lifetime is enough – if not too much.'

In a letter written much later, to Darwin's son, Huxley said: 'In justice to the Bishop, I am bound to say that he bore me no malice, but was always courtesy itself when we occasionally met in after years. Hooker and I walked away from the meeting together, and I remember saying to him that this experience had changed my opinion as to the practical value of the art of public speaking, and that from that time forth I should carefully cultivate it, and try to leave off hating it. I did the former, but never quite succeeded in the latter effort.'

Thirteen years later, on hearing of the death of Bishop Wilberforce, who had been pitched on to his head in a riding accident, Huxley commented: 'For once reality and his brain came into contact, and the result was fatal.'

> The ape is a degraded form that has descended from man. The donkey has descended from the horse.
>
> *George de Buffon (1707–88), the greatest naturalist of his day and author of a 44-volume account of natural science*

Who was G. W. Sleeper?

The history of science contains many cases in which permanent credit for major discoveries and ideas are attributed, not to their originators, but to the people who made the biggest splash in announcing them.

The names by which such matters are remembered often derive from the author of the book or paper that made the thing famous, such as Darwin's theory of evolution, Pasteur's germ theory of disease and Metchnikoff's theory of the phagocytes, but all of these may have been wrongly attributed.

Professor Sir Edward Bagner Poulton (1856–1943) was a scholar of Jesus College, Oxford who graduated with a First Class honours degree in zoology. He was an energetic man, President of the Oxford Union at the age of 23, a lecturer at Jesus and Keble colleges at 24, a Fellow of the Royal Society at 33, and Hope Professor of Zoology at Oxford at 37. He was President of the Linnean Society of London from 1912 to 1916 and became President of the British Association for the Advancement of Science in 1937. He was one of the foremost supporters of Darwin's theory at a time when there was still much opposition to it from the Church and others.

Something happened in 1914 that so deeply impressed Poulton that he made it the subject of his Presidential Address to the Linnean Society that year. A 36-page pamphlet by an American called G. W. Sleeper, printed in Boston in 1849, was sent to Alfred Russel Wallace (1823–1913), the co-discoverer of evolution, and Wallace sent it to Poulton. This pamphlet covered a lot of scientific ground. It put forward the idea that life on earth 'owes its faint beginning to primal germs . . .'. It suggests that all the species of today arose by diverging lines from a common ancestor; that 'man and the ape are co-descended from some primary type'. It proposes that infectious diseases are transmitted by invisible living organisms and that such transmission may be effected by insects. It describes the author's experiment in the isolation of streptococci from a sore throat. It puts forward proposals for the protection of buildings from insects with metal gauze window and door frames, and it puts forward the theory that germs in the body may be attacked by special scavenging cells that take them up and destroy them. Sleeper was even able to anticipate and use some of the terminology that was later to be used by the scientists who again made these discoveries.

Darwin published *The Origin of Species* in 1859, Louis Pasteur (1822–95) proved that infectious diseases were caused by micro-organisms around 1865, and Elie Metchnikoff (1845–1916) first proposed his phagocyte theory in 1883. Ronald Ross proved that malaria was transmitted by mosquitos on 20 August 1897. Yet there was clear evidence that, at some dates prior to 1849, a completely unknown American scientist or theoretical thinker had, quite independently of these great men, made all these wonderful discoveries. Amazing!

Professor Poulton was not in his dotage when he made this extraordinary revelation in his speech at the Linnean Society. He was a mere 57 and was to continue with his professorial work at Oxford until he was 77 and to be knighted two years later. He was a world authority on protective colouring in animals, and his book *The Colours of Animals* (1880) was a classic. His work had added a great deal of support to Darwin's theory of natural selection. So the professor was listened to with respect when he outlined the extraordinary detail with which G. W. Sleeper had anticipated so many later developments in science.

Most remarkable of all, perhaps, was the fact that those topics to which Sleeper had contributed just happened to be the most important scientific topics of the time. Darwin's work had changed the whole face of biology and had had enormous philosophical influence, Pasteur's work had changed the face of medicine and had saved countless lives, the importance of insect vectors as disease carriers was also a matter of the first importance, and the idea of the phagocyte was the first real breakthrough that was to lead to the fundamentally new discipline of immunology.

Poulton naturally addressed the question of the authenticity of Sleeper's pamphlet and, after considering all the possibilities, concluded that it was genuine. An editorial in *Nature* also showed due respect to the Professor's judgement. It concluded: '. . . but it may fairly be said that after weighing the interesting information brought together by Prof. Poulton respecting the book and its author, few will doubt that Mr Sleeper's work was really printed and published at the time stated, and that it contains one of the most remarkable anticipations of modern views and forms of expression respecting evolution and the germ-theory of disease that have yet come to light.'

All this is a little reminiscent of the revelations of the clairvoyant at a spiritualist seance. While purporting to foretell the future, these predictions always stop at a stage contemporary with that of the soothsayer. Moreover, they never contain anything that could not, somehow, be known to the seer. The most remarkable thing of all is that people of Poulton's and Wallace's scientific eminence should not have seen immediately that this was a barefaced fraud. It is unlikely that the editor of *Nature* was taken in, but these were still the days of authoritarianism when it was *de rigueur* to show proper respect for rank and status.

One can readily imagine what the great Scottish philosopher David Hume would have made of all this nonsense. 'Tell me now,' he would have said, 'which do you think the more probable? That this American person could really have done all these things, have expressed them in the language that was later to be used, and have kept quiet about them until now; or that he had simply changed the date on his outline of current scientific thought?'

To get this paper published by the Royal Society will require care and a little manoeuvring on my part. You have no notion of the intrigues that go on in this blessed world of science. Science is, I fear, no purer than any other region of human activity. Merit alone is very little good. For the last twenty years **** has been regarded as the greatest authority on the matter of this paper, and has had no one to tread on his heels, until at last, he has come to look on the Natural World as his special preserve. If my paper is referred to his judgement it will not be published. He won't be able to say a word against it, but he will pooh-pooh it to a dead certainty. So I must manoeuvre a little to get my poor memoire kept out of his hands.

Thomas Henry Huxley (1825–95), principal supporter of Darwin and grandfather of Julian and Aldous[1]

Piltdown

There was great excitement in archaeological and anthropological circles in 1912 when the fossil remains of a prehistoric man were found by the English solicitor, antiquarian and amateur palaeontologist Charles Dawson (1864–1916) in a shallow gravel formation at Barkham Manor, on Piltdown Common near Lewes in Sussex. Dawson, who was steward of the estate, was a real enthusiast who already had some quite important discoveries to his credit. He had

[1] The name signified by asterisks was that of Sir Richard Owen (1804–92), superintendent of the natural history department of the British Museum.

found the tracks of the megalosaurus dating back to the time when the Sussex rock was still mud, and he had discovered three new species of iguanodon, one of which had been named after him. He had also unearthed numerous fossil plants, fish and mammals. These achievements had brought the amateur scientist the coveted fellowship of the London Geological Society at the early age of 21.

In addition to his own finds, Dawson had briefed the local quarrymen at Hastings and elsewhere to look out for fossils, for which he was willing to pay handsomely. The result was that his house was soon overflowing, so a 'Dawson collection' was set up in the geology department of the Natural History Museum, a branch of the British Museum, in South Kensington, London.

One day in late 1911 Dawson was walking on Piltdown Common when one of a group of workmen digging gravel for farm roads offered him a piece of skull bone. They had, apparently, broken a skull while digging and had thrown it away. Dawson searched repeatedly and finally found two other fragments of the skull. These he took to the Natural History Museum and showed them to Dr Arthur Smith Woodward. Woodward, a most distinguished expert on fossil fishes and a Fellow of the Royal Society, was greatly excited and urged him to search further. But the winter rains had flooded the area and it was not until the end of May 1912 that Dawson was able to continue. Four months later, accompanied by the French Jesuit priest and paleontologist Pierre Teilhard de Chardin, he found the remainder of the skull bones and half of the lower jaw. Teilhard de Chardin also found a canine tooth that appeared to belong to the mandible. Associated with these relics were some bones of early animals and some crude flint implements.

Woodward was convinced that these relics were human. The skull closely resembled that of a young chimpanzee and did not have the massive bony protuberances over the eyes characteristic of a fully developed ape and of cave men. This suggested the theory that the skull changes occurring in the various stages of evolution of early humans exactly mirrored the changes that occur in the skull of an ape as it develops from youth to full maturity. Such a theory struck a resounding chord at the time, largely because a German palaeontologist and zoologist called Ernst Haeckel (1834–1919) had pointed out that the stages in the embryonic development of many animals closely resembled the evolutionary stages found in fossil animals (see Wishful thinking of a biologist). Woodward announced the extraordinary discovery at a meeting of the London Geological Society on 18 December 1912.

The news of Dawson's finds spread like wildfire around the scientific world for the skull was immediately deemed to be the most important archaeological discovery of all time. It was thought to be that of a woman, a representative of a race from which both the cave people and modern humans had sprung. Here, at last, was the missing link required by Darwin's theory of evolution. The fortunate discoverer found himself immortalized. His extraordinary find became known as 'Piltdown man' or *Eoanthropus dawsoni* – Dawson's man of the dawn.

Woodward drew some remarkable conclusions from Dawson's material. 'The most significant thing about this discovery,' he said, 'lies in the fact, proved beyond doubt from the shape of the jaw, that the creature, when alive, had not the power of speech. Therefore, in the evolution of the human species, the brain came first, and speech was a growth of a later age.'

Dawson died in 1916 and Woodward was knighted in 1924. Two years later a very strange thing happened. A detailed geological survey of the gravels at Piltdown showed that they were much more recent than had been supposed and that the dating did not correspond to the putative age of Dawson's specimens. Over the succeeding 40 years, other palaeontological discoveries, especially in Africa and Asia, raised further questions and made it increasingly hard to believe that the Piltdown findings could be accepted as what they seemed. Discoveries of more primitive samples of Australopithecus, and more samples of Neanderthal man, meant that Piltdown man no longer had a place in any plausible evolutionary sequence. Was it possible that the experts had blundered in their interpretation? Or was there another explanation?

Eventually, critical pressures rose to the point at which it became imperative to take another, much closer, look at the Piltdown remains. This intensive re-examination showed that they had been skilfully faked. The skull was, in archaeological terms, a mere thing of yesterday. It was certainly a human cranium, but it was no more than 50,000 years old. The jaw and the teeth were those of an orang-utan, and the single tooth probably from a chimpanzee. Even more remarkable, the animal remains that had been associated with the skull find could not have been of British origin.

Perhaps most damning of all, chemical analysis proved that the bone fragments had been stained, some with chromium compounds and some with iron sulphate. Since neither chromium nor iron sulphate was to be found naturally in the area, the presumption was that the staining had been done deliberately with intention to deceive. Closer examination of the teeth showed that they had been artificially abraded so as to produce a human type of wear pattern.

The mystery of the Piltdown man (or woman) remains unsolved to this day. It was certainly faked, but was it a hoax or was it a deliberate fraud? Who was responsible? Was Dawson hoping to become a Fellow of the Royal Society? Was Woodward ambitious for a coup of a lifetime? Was someone aiming to make Woodward look silly? Is it possible that Teilhard de Chardin, theologian and philosopher, was a party to the fraud? Could the whole disgraceful business have been a conspiracy involving all three men? What was the role of Sir Arthur Keith, anatomist and conservator of the Hunterian Museum of the Royal College of Surgeons, who allegedly provided technical expertise? Did he also provide the bones? Where did Sir Arthur Conan Doyle, the author of Sherlock Holmes, come into the picture? Conan Doyle lived near Piltdown, was an acquaintance of Dawson, and was interested in fossils.

All these questions have been asked by a succession of writers intrigued by the mystery and anxious to put forward their own hypotheses to account for it. None of them has ever been convincingly answered.

In animals, there is a gap between species which Nature cannot bridge. The Creator has dictated simple but beautiful laws that impress upon each species its immutable characters.

The French naturalist George Leclerc Buffon (1707–88),
denying the possibility of evolution

What killed the dinosaurs?

It is now widely accepted that the dinosaurs were wiped out by the impact of an enormous asteroid. This missile hit the earth 65 million years ago on what is now the Yucatán peninsula in Mexico, producing a crater 180 kilometres wide. The vast amount of pulverized material sent up such a massive dust cloud into the atmosphere that the darkness lasted for about three years. The absence of sunlight completely blocked photosynthesis and destroyed the food supply that was the basis of the food chain. So, along with hundreds of other species, the dinosaurs died out in a matter of weeks.

This theory was first proposed in 1980 by the American theoretical physicist Louis Walter Alvarez (1911–88). Alvarez was an unusual man with a remarkable range of interests. As a young man he developed the proton linear accelerator and the liquid hydrogen bubble chamber. These enabled him to discover a number of new subatomic particles and to win the 1968 Nobel Prize for physics. He measured the strength and direction of the magnetic field of the neutron, he was the first to show that muons could be used to bring about cold atomic fusion (see The cold fusion affair), and he proposed the implosion system that was adopted for the atomic bomb. His work on narrow-beam microwave radar led to the development of blind landing systems for aircraft. He also invented variable-focus spectacles, developed a radar system for aerial bombing, used cosmic rays to determine the presence or absence of concealed underground chambers in the Egyptian Chephren pyramid at Giza, and proved that President John F. Kennedy was murdered by only one person.

Alvarez's theory of the dinosaur extinction, developed in conjunction with his geologist son, Walter Alvarez, was greeted at first with a good deal of scepticism. But, as people got accustomed to the idea and assimilated the evidence, it gained wide acceptance. The Alvarezes showed that there was a worldwide thin layer of clay with an unusually high content of the element iridium. This is a feature of the rock strata deposited during the Cretaceous, the last period of the Mesozoic era, between the Jurassic and Tertiary periods. This layer, they suggested, had been deposited when the dust cloud from the asteroid impact finally settled. Iridium is rare on the earth's surface but is more plentiful in asteroids. So another possible source of the iridium deposit was provided. When, in 1991, radiocarbon dating placed the Yucatán catastrophe exactly at the right time – just at the boundary between the Cretaceous and the Tertiary – all doubts were resolved.

Or were they? Well, not among the scientists. Whereas the simplest explanations may, to the layperson, seem the most attractive, true explanations are not always the simplest explanations; there are, in fact, a number of other factors to be considered.

First is the fact that the dinosaurs did not all flourish right up to the time of the catastrophe and then all die out together. Some species of dinosaurs survived for as long as a million years after the presumed asteroid impact, and, according to some scientists, many species of dinosaur had been steadily declining for millions of years before the final curtain. The amount of dinosaur material available for study is strictly limited, but a study of the fossils of other living things, especially microfossils, shows that most of the extinctions of species began over a hundred thousand years before the time of the impact. An even more serious blow to the Alvarez theory is that many other species of animals, known to be highly sensitive to any environmental changes, lived through the impact period and went on to flourish.

Some experts have put forward an entirely different catastrophe theory. Astrophysicists are all agreed that stars are formed by the gravitational pulling together of clouds of gas. These clouds of cold gas – mainly hydrogen – are all over the universe and gradually aggregate by the normal forces of gravitation attraction that exist between all bodies. As the molecules are pulled together they get hotter – a phenomenon familiar to anyone who has blown up bicycle tyres. The process continues until the increased density, pressure and temperature at the core of the mass start up a nuclear fusion reaction (see The cold fusion affair).

But we run unnecessarily ahead. These experts suggest that, 64 million years ago, the earth passed through one of these early clouds of cold gas, which are known as giant molecular clouds. This passage could take a million years. Hydrogen will readily burn in oxygen to produce hydrogen oxide, better known as water (H_2O). No harm in that, you may think. But as lightning flashes repeatedly set fire to pockets of oxygen, bringing about this reaction, about a third of the oxygen in the atmosphere would have been combined into water and would not have been available to support life. Exit the dinosaurs.

Why should we believe this theory? Well, incredibly enough, we have evidence that the oxygen levels in air at this time were as low as the theory proposes. Bubbles of air preserved in amber from the period have been analysed and show the reduced oxygen concentration. In addition, there is the even more illuminating possibility that the earth may have passed through such clouds on a number of occasions, thus accounting for the other known, and previously unexplained, episodes of species extinction. Supporters of this theory claim that the atmospheric changes would be more dangerous to large species, such as dinosaurs, than to smaller creatures.

It is, however, unnecessary to offer extraterrestrial explanations for the demise of the dinosaurs or even for the other species extinctions. We know that the geological evolution of the earth was punctuated by massive volcanic eruptions: not just the odd volcano blowing its top but supervolcanoes affecting whole continents. The evidence that these actually did happen remains in the shape of enormous layers of basalt miles thick. Basalt can only come from the interior. So far as life on earth was concerned, these eruptions had a far more grave effect than just burning a few paws. They released into the atmosphere millions of tons of carbon dioxide and the horrible choking and toxic gas sulphur dioxide. The resulting combination of greenhouse effect with its temperature rise and the poison in the inhaled air would have had a devastating effect on animal life. The supervolcanism would also bring up from the earth's core plenty of iridium which would be nicely spread around.

There was a massive episode of supervolcanism about 250,000 years ago and it wiped out almost all living things. The end of the Cretaceous period was one of severe instability in the earth's crust and there was a considerable amount of volcanic activity. Some of the experts are convinced that it was this kind of thing, or perhaps a combination of supervolcanism and some of the other possibilities, that finished off the dinosaurs.

It is all speculative. Scientists are wonderfully human, just like the rest of us. The idea that they are engaged in a calmly disinterested search for truth and are totally unaffected by such unworthy considerations as personal pride, competitiveness and envy is a dream of adolescent idealism. The real world isn't like that. Like many successful scientists, Alvarez was a colourful and newsworthy character who attracted a good press. He also created more than a few enemies in the scientific community by his dismissive attitude to some people who had spent their lives patiently engaged in painstaking palaeontological research: he called them 'stamp collectors'. Perhaps most annoying of all was that, as an expert in physics, he had the temerity to wander into a totally different discipline and then achieve public recognition in it.

So it is not in the least surprising that Alvarez's asteroid theory met with opposition. It must have prompted many scientists to start thinking hard to come up with alternative hypotheses. It is a noted characteristic of human beings that when they do have a new and interesting idea that seems to them plausible, they will try to defend it against all comers. Facts which support the idea will be made much of and will prompt the search for collaborative detail. Facts which oppose it will not, of course, be ignored – that would be dishonest – but they will not be treated with the respect they deserve, and reasons will be sought to show why they can safely be dismissed.

The jury is still out on the cause of the death of the dinosaurs.

Physics

> Anyone who expects a source of power from the
> transformation of the atom is talking moonshine.
> *Ernest Rutherford (1871–1937), one of the great names in*
> *atomic physics*

Hero and the prime mover

The Greek scientist and engineer Hero of Alexandria who lived
some time in the first century AD, was a very ingenious man. He
wrote a textbook of engineering and was a prolific inventor of
machines.

His machine for opening temple doors is an excellent example of
his imaginative creativity. A large air-filled chamber is connected by a
pipe to the top of a cistern almost full of water. The whole system is
sealed except for a water pipe that runs up out of the cistern from a
point well below the surface of the water. When a fire is lit below the
air chamber the air is heated and expands, driving down the level of
the water in the cistern. The water that is forced out of the cistern falls
into a large bucket suspended from a rope that runs over a pulley. The
other end of this rope is wrapped several times round, and secured to,

a vertical pivot on which the temple door rotates. As the bucket fills, it descends, pulling on the rope and turning the door pivot. The whole of this mechanism, except for the fire, being invisible to the audience, the door would seem to open spontaneously in response to the lighting of the fire on the altar.

Hero was clearly aware of the power of expanding gases such as air and steam and developed several devices that could have been used to provide new sources of power and enormously expand human potential for wealth. Typical of these is an effective high-speed steam engine – a device that might have enriched humanity. This showed that Hero was familiar, in practice if not in theory, with the idea of action and reaction. It was, in fact, a primitive steam turbine. Hero's engine consisted of a metallic sphere pivoted at the two ends of a diameter so that it could rotate freely. Two pipes projected from the sphere at right angles to the pivotal axis. These pipes were bent in opposite directions. In one version the sphere contained water and had a fire under it to turn the water to steam; in an improved version steam was supplied to the sphere by a pipe from a boiler that entered the sphere via one of the pivots. When a sufficient head of steam had been obtained the escape of steam from the bent pipes caused the sphere to rotate rapidly.

This elementary prime mover was never, so far as we know, put to practical use, but instead was considered as no more than a toy. The extraordinary thing about all this is that Hero and his interested contemporaries – men such a Philo and Ctesibious – were completely familiar with a range of mechanical principles and devices, such as gears, levers, cams, pulleys and valves, that would have enabled them to apply these power sources in exactly the same way as they were applied seventeen hundred years later at the start of the industrial revolution. Hero even invented a screw-cutting machine that could have facilitated the development of even more complex machinery.

Exploitation of this knowledge and skill would have given the Romans, who then occupied Alexandria and ruled Egypt, and the Greeks, who studied in Alexandria, a powerful advantage over the rest of the world.

Was this a major scientific blunder or was it simply a reflection of the lack of wider imagination? The answer is as curious as it is unexpected. The answer was neither of these; it was plain snobbery. Doing things with your hands rather than with your mind was vulgar. Producing machines that could make life easier for other people was ignoble. It was perfectly all right to amuse yourself inventing these things and perhaps, even, writing about them. At the most, you might employ slaves to make your machines, just to see if they would work, but no gentleman would actually soil his hands with such work or encourage others to do so.

Benjamin Franklin's lightning conductor is a sacrilege that tries to avert the wrath of God. The destruction of Lisbon by the earthquake and tidal wave is God's punishment of man for this sacrilege.

From a sermon by a Boston minister in 1753

A galvanic error

The science of electricity can be said to have started with the ancients, who found that certain substances rubbed with wool or silk would attract very light objects. These substances are what we now call good insulators (poor conductors) and we know why they show this property. One of the best of these was the substance amber, which is fossil resin derived originally from coniferous trees. The Greek word for amber is *elektron*.

One of the first to make a systematic study of this phenomenon was a medical man called William Gilbert (1544–1603). Gilbert was President of the Royal College of Physicians, but he seemed to have plenty of time to devote to non-medical matters for he wrote a remarkable book about magnetism called *De magnete*, published in 1600. In this book Gilbert made the extraordinarily prescient conjecture that magnetism and the effect produced by rubbing amber were manifestations of the same force. He was, no doubt, influenced by the obvious similarity between the effect of a magnetic lodestone on iron and that of amber on little bits of paper, but there was more to his idea than he knew.

Gilbert did lots of trials and showed that many substances would work in this way. These substances he called 'electrics'. Other substances, which we would now call conductors, he entitled 'non-electrics'. Later, in 1646, another celebrated doctor, Sir Thomas Browne, used Gilbert's terminology to coin the term 'electricity'. Since conductors will convey electricity and insulators will not, Gilbert's idea of 'electrics' and 'non-electrics' may seem a bit odd. But this was not a blunder. The point is that only good insulators can readily build up a strong static electric charge so as to attract bodies of opposite charge (unlike charges attract; like charges repel). It's no good rubbing a rod of copper or silver. These are such good conductors that any induced charge is immediately conducted away onto your hand or the rubbing cloth.

Gilbert wrote voluminously and bequeathed all his papers, books and instruments to the Royal College of Physicians. Needless to say, they were all destroyed in the Great Fire of London in 1666, but his *De magnete* survived and was highly influential. For nearly 200 years he was just about the only authority of the dawning science of electricity and magnetism. The poet Dryden expressed the intellectual's view in an elegant iambic pentameter: 'Gilbert shall live till lodestones cease to draw.'

The next step in the history was made by an obscure amateur scientist, the dyer Stephen Gray (1666–1736). Gray's contribution was all but ignored at the time, but was most important. He actually succeeded in showing that electric charges could be conducted a distance of about 150 metres through a damp hempen thread (conductor) supported by dry silk strings cords (insulators) and, later, that charges could be even more effectively conducted through metal wire supported on insulators.

From about 1750 to 1820 or so, scientists were convinced that electricity was some kind of fluid. In 1733 the French chemist Charles François de Cisternay DuFay (1698–1739), stated that there were two kinds of electricity – positive or 'vitreous' and negative or 'resinous' electricity. He got this idea from the different behaviour of glass (vitreous) and amber (resinous) due to the difference in their insulating properties. DuFay also spoke truer than he knew, but didn't quite get it right. Electricity, he believed, was unquestionably a fluid. Ordinary objects had no electrical powers because they contained equal quantities of positive and negative electricity. But with some substances, rubbing caused separation of the two fluids, so that the object could then either attract or repel other light objects. This was remarkably close to the facts.

In 1746 a Dutch scientist called Pieter van Musschenbroek (1692–1761), who had been born and bred in Leiden, and who also was professor of physics at Leiden University, found a way of storing electric charges. This was the celebrated Leyden jar which was to provide an enormous stimulus to electrical research. Musschenbroek took an ordinary jam jar, cleaned and dried it thoroughly, and coated the inside bottom and halfway up the inside with tin foil. He then coated the whole of the lower half of the outside of the jar with foil. A waxed cork was fitted to the mouth of the jar and a brass rod ending in a short brass chain was pushed through the cork. The chain made contact with the interior foil and the brass rod had a brass knob on top.

The Leyden jar was, in fact, what was later to be called a condenser and is now called a capacitor. The foil formed the conducting plates and the glass the insulating dielectric. By means of various rubbing machines – such as those consisting of a rotating sphere of sulphur – large electric charges could be stored in the jar by way of the knob. Because unlike charges attract, a negative charge on the inner plate would induce an equal positive charge on the outer plate.

The Leyden jar was a great step forward and allowed scientists to carry out many electrical experiments and so advance their knowledge of the subject. By this stage in the progress of the science – we are now near the end of the eighteenth century – it was clearly understood that some substances, such as metals, could readily conduct electricity, while others, such as glass or amber, could not. The former were called conductors; the latter, insulators. It was also known that most bodies were electrically neutral and that the production of electricity on them involved the separation of positive and negative charges.

At about the same time as the invention of the Leyden jar, the brilliant American Benjamin Franklin (1706–90) was researching into electricity and made some important discoveries. By the hazardous expedient of flying a kite into thunderclouds, he was able to show that lightning was an electrical phenomenon. This was an extraordinary experiment. Franklin made a kite of cedar strips and a large silk handkerchief. To this he attached a sharp-pointed wire about a foot long. The twine string of the kite became wet and so was able to conduct electricity, but at the bottom end Franklin tied on a length of silk cord which he kept dry by standing in a doorway. He fastened a metal key at the bottom of the wet twine and, while flying the kite into a thundercloud, obtained sparks between the key and his knuckles.

Franklin also showed that electrical charges could readily be drawn off charged insulators by earthed metal points. So, in 1750, he developed the lightning conductor which was soon sprouting on most large buildings, including, in England, St Paul's Cathedral and Buckingham Palace.

This led to a great scientific controversy as to whether pointed or rounded lightning conductors were more efficient. Most of the members of the Royal society accepted Franklin's advocacy of pointed conductors but, at the time, Franklin was engaged in supporting the cause of American independence and was considered an enemy of England. This was enough to persuade some people to reject his views on everything. The king was talked into having all the palace points replaced by balls, and His Majesty tried to prevail on the President of the Royal Society, Sir John Pringle, to try to change the opinion of the members in favour of balls. It is greatly to the credit of Sir John that he immediately resigned.

In 1780 Luigi Galvani (1737–98) was professor of anatomy at the University of Bologna. Like most other lively-minded people of the time, Galvani was interested in electricity. As is often the case, it was the link between his own discipline – anatomy – and the new science that proved fruitful. Galvani decided to see what would happen when he applied electricity to the legs of recently dead frogs. The answer was a convulsive twitch. Galvani found that he could bring about this response in a variety of ways. The easiest way was to connect the leg with a metallic conductor to the knob of a charged Leyden jar. But he could obtain a twitch with no direct contact to an electrical device but merely by metallic contact to the muscle in the vicinity of a working electrical machine. He could even cause it by touching the muscle with the points of a pair of scissors during a thunderstorm. There was, of course, no way that Galvani could know that these effects were due to a current *induced* in the conductor, much in the manner that a radio aerial picks up a small current.

In the course of his work, Galvani made an even more remarkable discovery. This was that, with the machine not working, the Leyden jar discharged, and no lightning about, he could still get a twitch if he touched the leg with the free ends of pairs of dissimilar metal wires, such as brass and iron or silver and copper, joined at their other ends.

These strangely irreconcilable facts, especially the last, got Galvani thinking. After a time he came to the reasonable conclusion that, since on the basis of what he knew about electricity there couldn't possibly be any in the metals, the electricity must reside in the frog's muscles. Although this was actually a major blunder, it was by no means an implausible suggestion. At the time, and for a long time afterwards, there was a general belief among scientists in what they called a 'vital principle' that made the difference between inanimate and animate matter. If the '*vis vitalis*' (vital force) was present the thing was living; if it was absent, the thing was dead. What could seem more likely to Galvani than that electricity was the long-mysterious principle?

Galvani was in no hurry to publish his remarkable findings, but in 1791 he produced an essay on the subject that convinced most of his colleagues that he had made a notable breakthrough. At the University of Pavia in Lombardy, northern Italy, however, there was one scientist who was convinced that he was talking rubbish. This was Alessandro Volta (1745–1827). It is a well-known principle in science that when a person makes statements about a subject in which he is not professionally engaged, those who *are* officially the experts in the matter will make strenuous efforts to prove the amateur interloper wrong.

Whether or not this was the motive in the case of Volta, the fact is that he was a professor of physics and Galvani was a professor of anatomy. Electricity is unequivocally within the province of physics and has, on the face of it, nothing to do with anatomy. Volta insisted that the twitching was due simply to the effect of contact with dissimilar metals and that this was the source of the electricity. Soon the scientific community was divided into two rival camps and, for years, the arguments continued. Galvani was a tough fighter and came up with some very impressive experimental results as well as some very cogent arguments.

Ironically, although Galvani was wrong about the electricity residing in the living tissue, the truth is that muscle and nerve fibres (cells) do have a constant electrical charge of their own across the fibre membrane. Reversal of this charge is what brings about the flow of nerve impulses and the contraction of muscle fibres. None of that is, however, relevant to the present discussion. The fact is that Volta was right and Galvani was wrong.

Oddly enough, although Volta was convinced, and repeatedly stated, that the association of dissimilar metals could produce electricity, it took him years to come up with a convincing demonstration of the fact. Eventually, he was inspired to do so, but it was not until he had retired from his professorial chair that he got round to it. On 20 March 1800 he wrote a letter to the Royal Society in London describing a device that could produce a constant source of electricity. This was what was soon to be known as the voltaic pile. Volta took a thin disc of silver, about an inch in diameter, and placed on top of it a disc of zinc of the same dimensions. On top of that he placed a disc of thin leather that had been soaked in salty water. On top of the leather he then placed another silver disc, then a zinc disc, and another soaked leather washer. He went on adding discs in this exact order until he had a column of about 20 discs of each kind. This column was supported in a dry wooden frame and a wire was soldered to the lowest silver disc and another wire to the uppermost zinc disc. When these wires were touched a slight shock could be felt, and when the ends were brought close together a spark was produced. A few quick experiments showed that Volta's pile was certainly producing a substantial current of electricity.

The irony of the whole situation was that <u>Volta had no idea of the importance of his invention</u>. His purpose in producing it was simply to prove his point in the argument with Galvani and to show that Galvani was wrong. So much for the elevated motivation of the scientist. One of the most important inventions of all time was produced for reasons of personal pride.

Volta had a few problems with his pile, the most important of which was that his leather discs kept drying up, at which point the electricity ceased to flow. Volta thought about this and came up with an excellent solution. Instead of the piles of disc pairs he arranged a row of cups full of salty or slightly acidic water. Into each of these he inserted a small plate of copper and a small plate of zinc. Following the order in his pile, he used a wire to connect each copper plate in each cup to the zinc plate in the next cup. This equipment, known as the 'crown of cups', worked even better than the disc pile and produced electricity for much longer.

Some Fellows of the Royal Society and others to whom Volta had communicated his discovery now began to experiment on their own. They noticed something that Volta had also noticed, but had ignored: the copper or silver plated became corroded and the zinc plates were gradually eaten away. It was soon appreciated that this was important, that chemical changes were occurring, and that the electricity almost certainly came from these. Longer-term trials confirmed this and showed that the life of the primitive batteries could be equated with the life of the zinc and with the brightness of the surface of the other metal.

Soon an even more striking fact was discovered. When the free ends of wires from each end of the battery were dipped into salty water they soon became covered with tiny bubbles of gas. These gases could be collected in an upended test tube and it was not long before tests showed that the gas at one wire end was inflammable and that the other supported combustion. One was hydrogen and the other was oxygen – the two known constituents of water. It had been proved that electricity had the power to decompose water into its constituent elements.

All this was very hard luck for Galvani, who may well have been the better scientist of the two, but who, for the best of reasons, just happened to get the facts wrong. Galvani was well known in his day and his name has been perpetuated only in the crude electroplating known as 'galvanization', in which a coating of zinc is deposited on iron. Count Alessandro Volta, on the other hand has been immortalized in the daily use of his name for the unit of electrical potential – the volt.

Is there anyone anywhere such an idiot as to believe that there are people who stand with their feet opposite to ours, their legs in the air and their heads hanging down? Can there possibly be any place on earth where everything is upside down, where trees grow downwards and rain and snow fall upwards? These mad ideas are the result of the crazy notion that the world is round.

Firmianus Lactantius (fourth century AD), tutor to Crispus, son of Constantine the Great

---

Perpetual motion

There are plenty of examples of what appear to be perpetual motion around us – the rotation of the earth, the sun and the planets; the rotation of electrons round the nuclei of atoms; the continual motion of molecules, the effect of which can be observed under the microscope as Brownian movement; the extraordinary cycle of countless millions of tons of water that fall as rain and return to the atmosphere by evaporation; the endless changes in atmospheric pressure; the movements of the tides; and so on. But if we define our terms carefully, none of these can in fact be categorized as perpetual motion in the sense usually meant.

By that term we mean the continuing indefinite movement – usually rotation – of any device or object without the application of additional energy from an external source to sustain it. But the term means something more than that. People are not interested in the idea of perpetual motion so that they can sit and watch a wheel turning for ever. The fascination for the idea – the motive that has, over the centuries, driven hundreds of people to try to make such a machine – is the thought of obtaining a permanent source of free and unlimited power.

Changing atmospheric pressure can be used to move the corrugated and evacuated thin metal capsule of an aneroid barometer quite strongly. Such a device on a larger scale could do useful work. However, this is not perpetual motion, because the energy that causes the corrugated capsule to shorten and lengthen comes eventually from the sun. In the same way, the energy that causes the earth's water cycle – that drives, for instance, the electric generators on the Niagara Falls – also comes from the sun. The tides move by the gravitational pull of the moon and, in so doing, reduce infinitesimally the kinetic energy of our principal satellite.

Isaac Newton was the first to point out that, once a body is set in motion, there is no way that it will stop unless a force is applied to it to oppose its motion. A large and heavy flywheel, set up on bearings that involve the minimum of friction will, once it is set spinning, continue to rotate for hours. But there is no way that friction in the bearings can be wholly eliminated, and friction, however small, exerts a force that opposes the rotation of the wheel. Sooner or later the wheel will stop turning, and, of course, if the wheel is made to do work it will stop sooner. A spinning flywheel has kinetic energy and there is no way that this energy can be tapped for another purpose without diminishing the energy that keeps the wheel spinning.

Numerous claims have been made that perpetual motion machines have, in fact, been invented. These claims have been believed because descriptions and drawings of such machines often seem remarkably plausible. One of the earliest and most persuasive designs is that of a wheel with hinged arms each with a weight at its end. As the wheel is set turning the arms on one side fall outwards while those on the opposite side fold inwards. The effect, it would seem, is to produce an unbalanced situation that would cause the wheel to go on turning. The first known example of such a device was that designed by the thirteenth-century French Gothic architect Vilard de Honnecourt. Unfortunately for Vilard, the only unbalance the machine would have achieved was one with the lower weights hanging directly downwards so as to ensure that the wheel would not turn at all.

Following the expression of this idea, a number of increasingly ingenious ways were devised by other inventors for ensuring a similar apparent lack of balance on a wheel. One of the most impressive of these was designed and built by Edward, the 2nd Marquis of Worcester (1601–67). This machine, which the Marquis described in his book *Century of Inventions* (1663) was a wheel 14 feet in diameter with 14 spokes that curved out towards the edge and along each of which could roll a 50 pound ball of iron. The curve of the spokes was such that when the balls on one side were carried by gravity to the edge of the wheel, those on the other side were carried to its centre. This wonderful machine did run for an impressively long time. However, this was not due to the ingenious arrangement of the balls but to its weight and inertia. What the noble lord failed to note was that, although the position of the balls was certainly asymmetrical, so was the number of the balls on each side. There were always more balls on the side at which the balls were nearer to the centre, thus producing an exact balance. He also failed to notice that the machine ran somewhat longer without the balls, which inevitably contributed some friction. A virtually identical machine but using mercury instead of iron balls, was designed and tried, but met with no greater success.

Another clever idea was to arrange a heavy endless chain around a triangular section body with one long slope and one short slope. Since the section of chain on the long slope was longer and hence heavier than that on the short slope it must slide down. The chain would thus continue to rotate round the triangle until it wore out. Or would it? Regrettably, no. Undaunted, Sir William Congreve decided to improve on this idea. Instead of a chain he proposed an endless belt of sponge, so arranged that it was squeezed on one side of the slope but not on the other. The lower part of the sponge belt was dipped in water. The idea was that water would rise on the one side by normal capillary action but not on the other, thereby unbalancing the belt and causing it to rotate. No such luck.

A number of other designs of would-be perpetual motion machines made use of water. One of the simplest and most plausible-seeming consisted of a wide open-topped water-filled vessel with a narrow outlet pipe at the bottom. This pipe was curved round and up above the vessel so as to discharge down into it. The thinking behind this clever idea has persuaded millions. Because the weight of water in the wide vessel is so much greater than that in the narrow pipe, it must overbalance so as to produce a continuous circulation of water which could be used to drive a water wheel. Unfortunately for this theory, the weight of water at the outlet is equal only to that of the narrow column of water vertically above the outlet. This had been shown to be so, years before, by Simon Stevin (1548–1620) of Bruges, who demonstrated experimentally that the pressure exerted by a liquid depends only on its vertical height and not on the shape of the vessel in which it is contained. In fact, the pressure on the narrow outlet tube is simply that of the vertical column of liquid above it of cross-section equal to that of the narrow tube.

Another water-driven device was that designed in 1618 by the English physician, mystic and pantheist Robert Fludd (1574–1637). Fludd worked out that the energy produced by water passing over a mill wheel could be used to turn an Archimedean screw that would carry the water back up above the water wheel. This, of course, is simply yet another example of the way nature does not work. The way nature does *work* has been used by many inventors to simulate perpetual motion. But devices that rely on the forces of nature, such as changing atmospheric pressure, solar energy, or osmotic pressure, are not perpetual motion machines.

With each advance in science and technology, new ideas for achieving perpetual motion appeared. When Michael Faraday showed that electricity could be generated from magnetism and magnetism produced by electricity, a host of proposed schemes appeared in which an electrical generator produced a current that drove a motor that turned the generator. In principle, this is the same as the water wheel turning the Archimedean screw.

Coming to more modern times, some people decided that perpetual motion was here at last when they were told that superconducting electrical wires had zero resistance and could maintain a current flowing indefinitely. What they had forgotten was that superconductors cannot work at normal room temperatures and that a great deal of energy has to be spent in cooling them sufficiently to become superconducting. Much research is being devoted to the development of superconducting materials that will work at ever higher temperatures. We can safely assume that superconductivity at normal temperatures is a contradiction in terms.

The hope of achieving a perpetual motion machine has been encouraged by the production of a number of devices that appeared to achieve successful perpetual motion but that were, in fact, making use of natural sources of energy. As we have seen, changes of atmospheric pressure, for instance, operating an evacuated bellows of thin corrugated metal (an aneroid barometer) can be made to do work, such as lifting weights or winding up a clock. The energy used in this case is a tiny part of the vast amount that is constantly poured onto the earth from the sun. Sun energy can be derived more directly to heat water or to generate electricity from photocells and the heat and the electric current can be made to do work.

Even the energy from radioactive material can be made to do work as in the case of the mysterious radium clock of the distinguished British physicist Lord Rayleigh (1875–1947). In this device the leaves of a sealed gold-leaf electroscope separated and closed perpetually with no apparent source of energy. At the time, many even sceptical people were convinced that this was a genuine perpetual motion machine. We now know, of course, that the recurrent charges on the electroscope leaves came from the emission of charged particles from the radium as a result of spontaneous nuclear fission in which matter is converted to energy.

Perpetual motion has been a gold mine for fraudsters who have found it a wonderfully effective way of separating the gullible and the simple-minded from their money. Some of these charlatans have made a fortune promoting schemes to businessmen whose greed exceeded their common sense. The Philadelphian genius John W. Keeley (1837–98) invented a perpetual motion machine which he called the hydropneumatic pulsating vacuo-engine. This was demonstrated in New York before all-comers and attracted a great deal of capital, ostensibly for the purpose of developing the Keeley Perpetual Motion Motor Company. It was, of course, duly pocketed by Keeley. The machine was, in fact, powered inconspicuously from the cellar below it, but this did not become apparent until after Keeley's death. Perpetual motion charlatans can usually be recognized by the resentment they display towards anyone impertinent enough to insist on questioning their claims.

So is it fair to suggest that all these people who tried to produce perpetual motion machines were guilty of scientific blunders? Not entirely. The quest for perpetual motion, like alchemy and astrology, was a necessary stage in scientific development. Enthusiasts had to try these machines over and over again in order to be persuaded that they did not work. Then they had to think hard so as to decide why they didn't work. As they did so, new insights into science were obtained.

Ironically, the whole sciences of mechanics and hydrostatics are founded on the proposition that perpetual motion machines can never work. This realization also led to the formulation of the laws of conservation of energy and of thermodynamics – notable advances in scientific understanding. So why is it that perpetual motion machines can never work? To answer this question, we need to know what is meant by the law of conservation of energy. This is not nearly so difficult as it sounds.

When a force acts to make something move, it is said to do work. A force that doesn't move anything is not doing work. The heavy computer on your table is exerting a force downwards on the table but it is not doing any work. Motion apart from force also does no work. If you give an ice-skater a push you do work, but once the skater is in motion no work has to be done to keep him or her moving. If you can imagine perfectly frictionless ice stretching to infinity and no air resistance, the skater would just go on moving indefinitely at the speed you applied. This is just another way of stating Newton's first law of motion.

Anything that can cause something else to move – that is, that can do work – is said to possess energy. If you climb to the top of a hill, your muscles convert chemical energy from your food into movement (kinetic) energy and this is expended in getting you to the top. As you expend chemical and kinetic energy, you acquire an equal amount of energy – called potential energy – by virtue of your altitude. Step off a precipice and see whether your potential energy can do work. As you fall, you progressively lose potential energy but acquire an equal amount of kinetic energy. When you hit the ground, some of the kinetic energy is converted into heat and some of it is expended in bringing about a slight rearrangement of your body.

During the whole of this process, there is no loss or gain of total energy. Energy is conserved: it will be converted from one form to another, but will be neither created nor destroyed. A gallon of petrol contains an immense amount of chemical energy and this can do a great deal of work, as, for instance, in moving a heavy car 30 miles. This is a conversion of chemical to kinetic energy. The car engine and the exhaust get hot. This is a conversion of chemical to heat energy. If you drive uphill, some of it is also converted to potential energy. So we see that energy can readily be converted from one form to one or more other forms, but in the process the total amount of energy is unchanged.

Now consider a perpetual motion machine. According to Newton's first law, a force is needed to start it going, but no further force is needed to keep it going. However, no machine can work without friction. Friction produces heat, so energy is taken from the system and the machine slows down and eventually stops. The situation is even worse if you expect the machine to do work. It can only do so by exerting a force to move something. When it does this, it loses energy and slows down and stops. Bad luck for the inventor.

Today, now that the law of conservation of energy is know to all concerned, it need not occasion surprise that anyone attempting to obtain a patent for a perpetual motion machine will be sure to have their feelings hurt. Kindly but firmly, they will be informed that, however ingenious and convincing their plans may be, the Patent Office will have nothing to do with them.

The heavier a stone is, the faster it falls.
Aristotle (384–322 BC), Greek philosopher whose writings
influenced European thought for many centuries

Which way is down?

If you took an eight-inch iron cannon ball and a half-inch spherical bullet to the top of the Leaning Tower of Pisa and dropped them over the side you would probably be arrested. The idea of doing such a thing, however, would be to see which hit the ground first. Legend has it that Galileo (1564–1642) performed just such an experiment, although there is no record in his writings that he ever used the tower in this way.

Common sense might suggest that the cannon ball, being heavier, would fall faster, and this is what most people thought in the time of Galileo. Galileo knew the value of observation and experiment, and if he had performed this experiment in Pisa he would have found that the two objects would hit the ground at exactly the same time. To be strictly accurate, the small bullet would strike the ground first because, being smaller, the resistance offered to it by the air would be less. However, this effect would be too small to observe in such a crude trial.

Most people, having been conditioned to believe that you can't always trust common sense in matters scientific, would be quite happy to accept that the two bodies fall at the same speed. In addition, they would be further convinced by the standard demonstration of a feather and a coin falling at the same speed in a long glass tube from which the air has been sucked out to remove air resistance.

However, those who are familiar with Newton's law of gravitational attraction may be puzzled: the force of attraction between two bodies is proportional to the two masses multiplied together. So, the cannon ball, being of larger mass, and being attracted by the earth, will have a greater force pulling it, and a greater force will make it go faster!

In fact, the acceleration caused in the two bodies is the same. Acceleration of a body produced by gravity is independent of the mass. But why is this so if the force acting on the larger mass is greater? The answer is actually quite simple. Let's suppose that we use a drop of SuperGlue to fix a string to the cannon ball and another drop to fix another string to the bullet. Now we hold the other ends of the two strings and think of ourselves as the earth pulling the two objects towards us by gravitation. Which is harder to pull? Why is the cannon ball harder to pull? Because of its mass, of course. Why is it harder to start it moving? Inertia. Heavier bodies have greater inertial mass and require greater force to accelerate them. Now, gravitational mass is exactly equal to inertial mass, which means that the greater force of gravitational attraction acting on a larger mass is exactly equivalent to the greater force required to move it. Hence the cannon ball and the bullet will experience the same acceleration.

It may be difficult to envisage why there should be a problem starting the movement of a falling body, especially a heavy falling body like a cannon ball. Falling implies a *downward* movement, and this raises the question: 'Which way is down?' Someone in London knows which way is down, but wonders which way is down for someone living in Australia.

Imagine you are an astronaut and have just left the moon on the return trip to the earth. The distant earth is seen to be spherical and, in relation to your space vehicle, you can arbitrarily imagine that the earth has a top and a bottom. As you get closer and closer to the earth, the sphere comes almost to fill your whole field of vision. At that point you can see only a hemisphere and you begin to get the feeling that the middle of it is below you. The closer you get, the stronger is this sense of the surface being 'downwards', and by the time you are orbiting the earth there is no doubt about it in your mind. We also 'feel' which way is down because we have gravitationally sensitive sensors in our inner ears. So our sense of what is down is partly psychological, partly physiological and partly just logical.

Now consider that you are on another trip, this time to a tiny asteroid only 50 feet or so in diameter. When you reach it and go on a space walk to have a look at it, you may have problems in deciding which way is down. The asteroid's gravitational pull on you is so small that you just drift around it. But there *is* a pull and, if you wait long enough, you will land on it. You may approach the surface from any direction and land anywhere on the surface. Every one of these 'downward' routes is 'down'.

So 'down' is not a particular direction. It is the direction between a body of small mass and the centre of the nearest body of large mass. The force required to overcome the inertia of the cannon ball falling to earth is the same regardless of how its position appears from a distant spacecraft.

> Light travels through water faster than through empty space.
>
> *Isaac Newton (1642–1727), the greatest scientist and mathematician of the seventeenth century*

Newton got it wrong

Isaac Newton was one of the few really great scientists of all time. His law of gravitation dominated physics for 240 years and was the basis for many remarkable advances in the understanding of science. It has been described as one of the most important advances in all scientific understanding and, in a sense, this is true. The only snag is that it is wrong. This is not to say that Newton was a blunderer – far from it. But in taking up the difficult subject of gravitation he fell into an error so fundamental that when it was eventually corrected, by Albert Einstein, the whole face of science was changed.

The Newtonian law of gravitation states that the gravitational attraction between any two bodies in the universe is directly proportional to the two masses multiplied together and inversely proportional to the square of their distance apart. Alternatively, you can say that the attraction is *equal* to a constant multiplied by the two masses multiplied together, the result being divided by the distance apart multiplied by itself. For example, if you have two bodies, each of mass 10 pounds, the attraction between them is proportional to 100 pounds and will thus be about ten times the attraction between two bodies of 3.33 pounds mass, the same distance apart. If two bodies are a foot apart and you then move them to three feet apart, the attraction between them will be reduced, not to one-third, but to one-ninth – the square of the distance.

If you move them to 10 feet apart, the attraction will be reduced to one-hundredth. The constant, called the gravitational constant, or *G*, is very small and its units depend on which system you choose to use for the masses and the distance.

The weight of a body is the resultant effect on it of all the gravitational forces of the universe. In terrestrial practice, however, the weight is, for practical purposes, the result of the gravitational pull of the earth. The mass of all other terrestrial bodies is negligible compared with the mass of the earth. A body *weighs* much less on the moon than on earth, because of the lesser gravitational pull, but the *mass* of that body doesn't change. A body at rest can be made to move, that is, can be accelerated, only by applying a force. The more massive the body, the greater the force required to accelerate it. In other words, for a particular body, the ratio of the force applied to it to the acceleration that results from applying the force is constant. That ratio is what we mean by mass. This is why the mass of a body is the same on the moon as it is on the earth. The weight of a body is the force acting on that body, and this will be significantly different on the earth and on the moon. This is the subtle difference between weight and mass.

Incidentally, the word 'acceleration' means a good deal more than just 'getting faster'. To the physicist, acceleration means any change in speed or direction. Thus, the earth moving in its orbit around the sun is experiencing a constant acceleration. The only kind of movement that does not involve acceleration is a constant speed in a straight line. If you are travelling at an absolutely constant speed, in a dead straight line, in a car with closed blinds on a perfectly smooth road, there is no way you can tell you are moving. But if the car experiences an acceleration, by slowing or speeding up, or turning, or even hitting a bump, you will know it at once.

All this works very well in most circumstances. Newtonian physics is remarkably accurate and has served us very well for nearly all practical purposes. But it is not accurate under certain physical conditions, and there are questions that it could never answer. To Euclid and Newton, space did not affect motion. Events occurred in space but were not influenced by it. This is not correct. Events and space are inextricably interrelated because the nature of space causes events, such as movements of bodies, and events, such as the movement (or even the presence) of bodies, alter space.

We saw that mass was defined as the ratio of an applied force to the resulting acceleration. Well, the fact is that acceleration and gravitation *are the same thing*. Einstein's general theory of relativity, which he produced several years after the special theory, deals beautifully with the idea of gravitation and answers a number of previously unanswerable questions. This theory does away with the idea of a gravitational *force* altogether. Wherever there are bodies with mass we must think in terms of Einsteinian space-time, a four-dimensional amalgam of space and time.

The geometry of space-time is not Euclidean and it is far from uniform; it is a very bumpy curved surface (a geodesic) covered with deep and shallow hollows. Objects of great mass lie at the bottom of a deep hollow in space-time and bodies of smaller mass sit in shallower hollows. The earth being a body of great mass, sits at the bottom of a very deep bowl out of which it is hard to get. In fact, a body has to be accelerated to about 24,000 miles per hour to get out. All bodies with mass are creatures of space-time subject to its geodesic. A falling body behaves in a manner determined by the nature of its local space-time. It follows the easiest course from place to place. Acceleration occurs because the local effect of the body on space is to distort it so that the body is, as it were, up on the slope of a hollow down which it naturally rolls.

What applies to bodies applies to all electromagnetic radiation including light. Light is still travelling in straight lines, but, because mass causes space to be curved, the light appears to be deflected. The space-time geodesic of the earth, which light must necessarily move in, is such that a fine beam of laser light would curve downwards about 1 millimetre in 5000 kilometres.

Four-dimensional space-time is hard to visualize but analogies might help. Start with imagining a universe consisting of the surface of a large sphere of finite radius. This universe is finite and is inhabited by flat two-dimensional creatures who are necessarily confined to the surface of the sphere and are able only to slide around on it. To the flatties, the universe is unbounded. Flatties can go as far as they like in any direction and they can't ever come to the edge of their universe, though they can, of course, if they travel far enough, get back to their starting point. If a very intelligent flatty was thinking about its universe, it might have the hare-brained notion of postulating another dimension. All its friends would say it was mad because, being two-dimensional creatures, there was no way any of them could imagine a three-dimensional universe.

But we can. However, we can sympathize with their difficulty because what we have to do, being three-dimensional creatures, is to try to imagine a four-dimensional universe. This is what our universe really is. It is a four-dimensional space-time continuum, containing a finite amount of space, but it is unbounded. Like the flatties, we can travel as far as we like in any direction without ever finding an edge to the universe. (See also Euclid got it wrong.)

Most of us still consider space and time to be completely different entities. That view is acceptable for most ordinary purposes, but it will not do for all purposes. It is actually wrong and misleading. There are several different systems for specifying a point in space, all of them requiring three quantities (or coordinates). But we have been ac-customed to thinking that there is only one way of indicating a point in time – stating how long it was after some fixed event. Unfortunately for this simple idea, different observers have different perceptions of simultaneity. One may perceive two events, A and B, as being simultaneous, a second may see A occurring before B, while a third may perceive A occurring after B. This is not a matter of inaccuracy but an inescapable consequence of differences in the position of the observers and the fact that <u>information cannot be conveyed faster than the speed of light</u>. Things get even more uncertain if the observers are moving relative to the events perceived or vice versa, or both.

Let's consider that event B is a flash of light sent out in all directions from point B. Event A is a radio signal from a source A situated a long way from point B. The idea is to try to send the flash and the signal simultaneously. How is this to be done? It's no good having a photocell at A that triggers the radio signal when it receives the flash from B. This will only ensure that A comes after B. Light travels fast but still takes a finite time to travel anywhere. It's also no good having a radio receiver at B that triggers the light flash when it receives the signal from A. This will only ensure that B comes after A. Radio signals travel at the speed of light.

Right. Let's provide both A and B with amazingly accurate atomic clocks, perfectly synchronized to send out the flash and the signal at the same moment. Now we have achieved simultaneity. But how will anyone remote from the two sources know this? They can only know it by knowing that they happen to be at rest relative to the sources and situated at precisely the same distance from both of them. Now this is a very artificial and limiting situation, and the chances are that the two observers are not only at different distances from the two sources but also that they have no idea how they are located relative to them or even how they are moving relative to them.

These niceties don't matter too much in everyday earthly practical affairs in which light can be thought to travel instantaneously. But they must, for instance, greatly concern astronomers. A flash of light from our sun takes about eight minutes to reach the earth and hundreds of years to reach the inhabitants of a planet in a distant solar system, and there is no one fixed point from which different positions and speeds can be judged. All motion is relative, so you can take any point as your reference point and judge everything else in relation to that.

For reasons such as these, it is not possible to separate space and time. Each affects the other and they have to be considered together. But the idea that time is a dimension comparable to the three we are familiar with is still hard to take. Maybe another analogy will help. Illustrators commonly produce on a two-dimensional surface, such as a sheet of paper, a representation of three dimensions. They draw a cube in perspective and everyone knows what it is. Now, suppose we want to produce in a three-dimensional frame a representation of four dimensions. How do we do it? Well, we start with a three-dimensional figure – say a real cube. One way we can represent, diagrammatically, a further dimension is to have a number of identical cubes overlapping the first and each other, or even superimposed on each other. Our perception of these multiple cubes involves successive observations, and these involve time. This is how we make observations in the real world.

An approximation to true Euclidean space exists only when we are a very long way from bodies with mass. To find it, you would have to find a spot a long way from any stars. For practical purposes, of course, when we are concerned with the kind of relatively small areas with which humans are normally concerned, the approximation to Euclidean space is quite good enough. Space can be considered flat and Euclid rules. But outside everyday affairs – when we are dealing, for instance, with cosmic matters – Euclid will not do.

Newton was working on the assumption that Euclidean space was everywhere. Newton was wrong. This was, however, an honest mistake and he did not have the information necessary to prompt him to correct it. So far as gravitation was concerned, Newton never claimed to have any idea of its cause. In a letter to his friend Richard Bentley, he wrote: 'You sometimes speak of gravity as essential and inherent to matter. Pray do not ascribe that notion to me; for the cause of gravity is what I do not pretend to know, and therefore would take more time to consider of it . . . Gravity must be caused by some agent acting constantly according to certain laws; but whether this agent be material or immaterial I have left to the consideration of my readers.'

Trust you will avoid the gigantic mistake of alternating current.

Lord Kelvin (1824–1907),
writing to Niagara Falls Power Company

The great AC/DC blunder

The presiding genius over the science and technology of electricity was the English physicist and chemist Michael Faraday (1791–1867). Faraday, the son of a blacksmith, was one of 10 children and had little schooling. The family fell on bad times and came near to starving. For a time they were dependent on public relief under the poor law.

At the age of 13, Faraday was taken on as an errand-boy by a bookseller and bookbinder called George Riebau, who had a shop in

Blandford Street, London. Faraday did well as a messenger and after a year was offered an apprenticeship as a bookbinder. He had a very lively mind and was insatiably curious and this job suited him very well. So, although a conscientious worker and busily engaged with the sewing frame, boards, leather and gold leaf, he was able to spend much of his spare time voraciously reading the books he was binding. Those on science interested him specially and he kept notebooks in which he jotted down anything scientific that caught his attention.

One of his customers was so impressed by his knowledge that he gave him a pass to attend a course of lectures by Humphry Davy (1778–1829) at the Royal Institution. Faraday took detailed and accurate notes and, at the end of the course, copied these out neatly, made coloured drawings and bound them in leather. This book he sent to Sir Joseph Banks, President of the Royal Society, with a polite request that he should be considered for a job in the Institution. Banks ignored him. So Faraday made another set which he sent to Davy. As a result, at the age of 21, he was offered a job. The pay was less than he had been earning as a bookbinder, but he accepted gladly and the research laboratory became his new home. From the start, he worked furiously, putting in long hours.

At first he was treated as a mere janitor. All he was allowed to do was to wash bottles, clean the benches, sweep the floors and generally hump and carry. But Faraday watched everything that was going on and already had a creditable knowledge of science. Gradually, and very tactfully, he began to make deferential suggestions. Most of these were so sensible and helpful that, eventually, Sir Humphry allowed him to participate actively in the work. Soon he was appointed an assistant technician and was absorbed in research. In no time he had become indispensable to Davy, who even took him with him on a lecture tour of the leading cities of Europe.

Lady Davy was a silly snobbish woman, who was incapable of

appreciating Faraday's quality and did not share her husband's growing high opinion of him. As far as she was concerned, Faraday was a servant of a lower social order. She had, unfortunately, forgotten that her own father-in-law was a humble Cornish woodcarver. On the tour, Lady Davy insisted that Faraday should be treated as a servant, and she never missed an opportunity to show him his proper place. Faraday wrote to a friend: 'She likes to show her authority, and I find her extremely earnest in mortifying me.'

Her conduct reached a climax when they reached Geneva. There, a distinguished scientist, Professor de la Rive, invited the Davy entourage to dinner. A place was set for Faraday, whose status as a scientist had already become known to the professor. When they entered the dining room, Lady Davy objected to the presence of Faraday at table and insisted that he should have to eat with the other servants. The professor froze. He then turned to his butler. 'Set a place and serve dinner for M. Faraday in a separate room,' he ordered, 'See that he is served with all the dignity befitting a lonely young natural philosopher who is living on a plane high above the petty quarrelling of his associates.'

Until the age of 30, Faraday devoted himself almost exclusively to chemistry, which was Davy's discipline and which had brought him a European reputation. But now Faraday's interest in electricity was kindled and it was not long before the young man's brilliance began to overshadow that of his master. Davy, who had long enjoyed the position as a doyen of science, began to become a little jealous. Faraday had read a paper by the Danish physicist H.C. Oersted (1777–1851), who had discovered that a magnetic compass needle was deflected by an electric current flowing in a nearby wire. This set Faraday thinking and prompted him to construct a device that could produce a continuous rotary movement of a wire in a pool of mercury. This was the first demonstration of the conversion of electricity into continuous movement and, although crude, was the first electric motor. Davy, who had tried to achieve something similar but had failed, was furious and accused Faraday of stealing his ideas. Faraday calmly asserted that the idea was his own. When, his fame spreading, he was put up for membership of the Royal Society – a high honour for a young man – Sir Humphry, who was by then President, tried to block his election. Fortunately, there were others who knew of Faraday's quality, and who were also aware of Davy's *hubris*, and the Chairman was overruled.

In 1825 Faraday was made Director of the Institution laboratory. But he was much more concerned with scientific problems than with personal ambition or reputation. In the summer of 1831 he carried out a remarkable experiment that was to have profound consequences. He took a soft iron ring, about 6 inches in diameter, and wound two coils of wire round it, these being entirely separate from each other and insulated from the iron ring. He called one side A and the other B. The ends of the B coil he took to a wire passing closely over a magnetic needle, and the ends of the separate A coil he connected to his battery. As soon as he made the contact, the magnet needle swung to one side and then slowly back again to its original position. When he broke the contact, the needle swung once again.

Not only had Faraday invented the electrical transformer, but he had done something even more important: he had proved the existence of electromagnetic induction. At first Faraday could not understand why the needle did not show a constant deflection so long as the current was flowing in the first coil. The needle always moved on making and on breaking the contact in the A coil, so it was clear that a current was being induced into the B coil at these times, but it seemed that no current flowed in coil B when a steady current flowed in coil A.

Still puzzled, Faraday then carried out, for the first time, the experiment of pushing a magnet into a coil of wire connected to a galvanometer – a sensitive electric meter. The meter showed a deflection when the magnet was moving but not when it was still. Clearly movement was essential. Now that he knew the principle, the application was straightforward. Electricity could be produced by moving a magnet near a conductor or a conductor near a magnet. It was now merely a matter of developing machines to do this more efficiently.

Faraday had invented the two major devices that were to be the entire basis of the electricity industry – the electric generator and the transformer. These were by far the most important contributions ever made to the technology of electricity, and the world was never the same again.

At first, electricity was used largely for signalling by the electric telegraph. This went ahead rapidly. Only one example of many is that of William Thomas Henley (1814–83). Henley rented a room in Whitechapel, London, in 1836 and turned his old lathe into a wire-wrapping machine to produce insulated wire. He was wrapping wire with silk or cotton. Soon he had to move twice into larger premises and was employing 23 workmen. In 1859 he took 3 ½ acres of land at Woolwich and spent £7000 on building a factory to make long-distance cables.

Between 1868 and 1873 Henley manufactured 12,000 miles of submarine cable. By 1873 the land covered by his works had increased to 13 acres and he had three steamships and a 400-foot wharf on the Thames. He had spent £500,000 on the Woolwich works and his net profits amounted to £560,000.

The real gist of this section is the generation of electricity for lighting. Humphry Davy had noticed the spark between the terminals of an electric cell and went on to show that a much better light could be produced by causing a spark between pieces of hard carbon. For a time, the carbon arc was the principal source of electric light, and, with developments in Faraday's generator, powerful arc lights could be produced. These were urgently needed for lighthouses and were soon being used for all kinds of purposes. In 1878 an evening football match was played at Bramall Lane football ground in Sheffield under such lights. Arc lights were also used in theatres and were later to be used in cinemas.

The real explosion of the electric lighting industry did not occur, however, until the incandescent lamp was invented by Joseph Swann (1828–1914) and developed by Thomas Alva Edison (1847–1931) and others. Arc lamps were fine for industrial purposes and large-scale installations, but were quite unsuitable for domestic use in the home. Edison's first incandescent electric lamp, produced in 1879, had a filament of carbonized cardboard connected to platinum wires sealed into the glass of the base of the bulb. There was a 'pip' on the top through which the air was sucked before sealing. The electric supply was connected to two bare terminals on the wooden base into which the bulb was set. Carbon filaments were, at first, modestly successful. Edison made a filament of carbonized cotton thread for a lamp that burned for 40 hours. Many experiments were to follow, however, to find the ideal material for filaments.

Faraday's invention of the electricity generator was quickly taken up by others. The most important of these was the German electrical engineer Werner von Siemens (1816–92). Siemens was the eldest of four brothers, three of whom had distinguished careers in engineering and one of whom did so well in England that he decided to become a naturalized British citizen and ended up as a knight. Werner had been an artillery officer in the Prussian army and had developed a telegraph system. It was he who discovered that gutta percha was an excellent insulator for wire. In 1874 he set up the firm of Siemens & Halske, which became one of the world's greatest producers of electrical engineering equipment. But Werner's main claim to fame was his invention of the dynamo principle.

The trouble with the earliest generators was that they had to depend on permanent steel magnets which were weak and tended to lose their magnetism. The weakness of the permanent magnets of the day seriously limited their electrical output. Werner had read Faraday's book *Experimental Researches in Electricity* (1839–55) and knew that a powerful magnet could be produced by winding a coil of wire round a soft iron core and passing a current through the wire. So it occurred to him that this might possibly be the solution to the weak magnet problem.

The upshot of this idea was a communication read by his brother William to the Royal Society on 14 February 1867. It began: 'Since the great discovery of magnetic electricity by Faraday in 1830, electricians have had recourse to mechanical force for the production of their most powerful effects; but the power of the magneto-electric machine seems to depend in an equal measure upon the force expended, on the one hand, and upon permanent magnetism on the other. An experiment, however, has been lately suggested to me by my brother Dr Werner Siemens of Berlin, which proves that permanent magnetism is not requisite in order to convert mechanical into electrical force; and the result obtained by this experiment is remarkable, not only because it demonstrates this hitherto unrecognized fact, but also because it provides a simple means of producing very powerful electrical effects.'

Werner's idea was to replace the permanent magnet with electromagnets and to energize these with electric currents derived from the rotating coil (the armature). When the armature is first turned, only a very small current is produced, but this current flows through the electromagnet coils – now known as the field coils. As this current flows the soft iron cores of the electromagnets are magnetized. The result is a stronger current from the armature. This, in turn, leads to stronger magnetization of the field. Siemens called his new machine the dynamo-electric machine, the term being derived from the Greek *dynamis*, meaning 'power'. This was too much of a mouthful and was soon contracted to 'dynamo'.

Werner's brother went on the describe how, as the machine was turned, the resistance to turning became ever greater until either the driving belts slipped or the coils became so hot that the insulation on the wiring caught fire. This was wonderful news. The increased resistance was simply a reflection of the fact that, by the principle of conservation of energy, the enormously greater output of electrical energy required a proportionately greater force to turn the machine.

Perhaps unsurprisingly, this same idea had occurred to two British electrical scientists – Professor Charles Wheatstone and a Mr Alfred Varley. Wheatstone actually read a paper to the Royal Society on the same day as Siemens and Varley had applied for a patent in December of the previous year. Someone had to think of this fairly obvious development and the time was ripe for it. Simultaneous discoveries of this kind are not uncommon in science and technology.

Prior to the invention of the mechanical generator, electricity had been produced by rather crude chemical cells or by voltaic piles – columns of discs of alternating dissimilar metals separated by discs of cloth soaked in weak acid. These cells were of little use for serious long-term production of current but, in the eyes of the scientists and engineers of the time, they had one great advantage – they produced a steady direct current (DC), although, admittedly, one that soon faded off to nothing. The earliest dynamos, on the other hand, produced a current that rose to a peak in the positive direction then fell to zero, then rose to a peak in the negative direction then fell to zero again, and so on. This was alternating current (AC) and the number of such cycles per second depended on the speed with which the generator armature was turned.

Alternating current was at first widely used, and it was, if anything, better for arc lamps than direct current. Soon, however, a strong prejudice grew up against alternating current. There were various reasons for this. The early electric motors would run only on direct current. Electric storage batteries could be charged only by direct current, and direct current was essential for electroplating. Storage batteries were very popular in the early days of electricity. They allowed generators to be shut down for maintenance and repairs, and they provided a standby service in the event of generator failure. For these and other reasons, the design of the second generation of early electrical generators was concentrated on DC machines, and domestic and industrial electricity supplies were almost exclusively DC.

This was a serious mistake as time was to show. Direct current systems persisted until well into the twentieth century – in some cases until about 1950. But, in the end, all these systems had to be expensively scrapped and replaced with AC generators and distribution systems. Why was this?

As usual, it is easy to be wise after the event. Today we are well aware of the advantages of AC over DC and in those days they were not. But this was still a blunder. Faraday had invented the transformer and all the knowledge existed to make full use of AC as we do today. Some of the pioneers must also have been aware of the likelihood that distribution of electricity would soon extend from a very local affair to a much more widespread system. Moreover, they must have recognized the immense advantage of being able to change voltage at any time.

Be that as it may, it took a long time for the engineers to get round to changing over to an AC system. No doubt long before that happened many of them had regretted the earlier decision. Alternating current at a frequency of 50 or 60 hertz (cycles per second) produces no flicker in incandescent lamps. It allows the use of transformers of all sizes from the massive to the miniature, whereas direct current does not. Transformers consist of two or more separate and unconnected coils of wire wound on a common metal core. When an alternating current flows through one winding, an alternating magnetic field is induced in the core and this induces an alternating current in the other winding or windings. The different windings are insulated from each other. The main point, however, is that the voltage across the ends of a secondary winding may be higher or lower than that applied to the primary winding. The difference simply depends on the ratio of the number of turns on the two windings.

This means that AC-generated electric power can be distributed at very high voltage and comparatively low current. The advantage of this is that wires of smaller cross-section can carry the same power, or, more significantly, the same wires can carry much more power without overheating, as it is the current that causes the heating. In a DC system it would require massively and impossibly thick conductors to carry the currents required by modern societies.

Large modern AC generators in power stations produce current at 25,000 volts. This is fed to transformers that step it up to 400,000 volts for distribution by the national grid. From the pylons, conductors carry the current to substations where transformers reduce the voltage to 125,000 volts. This current is further distributed to heavy industry and railways, etc where, before use, transformers bring the voltage down to 33,000 volts. Further substations have transformers that cut the voltage to 415 for light industrial purposes and to 240 or 120 volts for domestic use. With every drop in voltage the available current rises. A system of this kind, operating on DC, is inconceivable.

Alternating current offers many advantages even at a domestic level. Mains AC enters the sealed power units of personal computers at 240 volts and is converted by a transformer to 12 and 5 volts, and then 'rectified' to DC and smoothed for the electronic circuits. Radio and TV receivers, hi-fi equipment, doorbells and other domestic devices also require transformers to adjust the voltage appropriately. The frequency of the AC generators is kept very accurately to the standard number of cycles per second. This allows electric clocks to be produced that keep good time. This application of alternating current has now, however, been overtaken by cheap, battery-operated electronic quartz movements which achieve an even higher standard of accuracy. Back to DC.

It is apparent to me that the possibilities of the aeroplane have been exhausted.
Thomas Alva Edison (1847–1931), one of the great inventors of all time and holder of more than 1000 patents

---

The heat fallacies

In everyday language it is convenient to talk of heat and cold as if they were two different things. Indeed, until the end of the eighteenth century they *were* considered different and it was widely believed that cold could pass from one object to another. In 1791, however, the Swiss physicist Pierre Prévost (1751–1839), professor of physics at the University of Geneva, finally debunked this idea. It was, he showed, the absence of heat rather than the acquisition of cold that caused the unpleasant sensation.

Prévost was a clear-thinking man and went further. He suggested that all bodies, even very cold bodies, gave off heat at all times. He also proposed that even bodies whose temperature was constant were radiating heat, and explained this paradox by suggesting that they were, at the same time, receiving an equal amount of heat from their surroundings. These were remarkable propositions for his time and were subsequently shown to be correct.

At that time heat was considered to be a kind of fluid and was called caloric. But in 1798 the American physicist Benjamin Thompson, Count Rumford (1753–1814) decided to go and watch some canon barrels being bored using the power provided by two horses. This experience was to have momentous consequences. As Rumford watched the slow noisy turning of the drills, he was astonished at the amount of heat produced. According to current belief, the caloric was released from the iron as the metal shavings were separated. So Rumford decided to make a rough estimate of the amount of heat produced, and, when he did, it immediately became obvious that there was something very wrong with the caloric theory.

This set Rumford thinking and stimulated his interest in experiment as a source of knowledge. The following winter Rumford carefully weighed a quantity of water, let it freeze solid, and then weighed it again. There was no difference in the weight. Since 'caloric' could be very light in weight, Rumford used a balance that was sensitive to one part in 1,000,000, but, whether frozen or liquid, the weight was exactly the same. This was another blow to the caloric theory. Thinking back to the heat produced in boring canon, Rumford concluded that the movement of the canon borer must be the source of the heat. Heat, he decided, was a form of motion.

Rumford took the matter further and actually measured the amount of heat produced by boring canons. He did this by surrounding the canon barrel with a box full of water and measuring the temperature rise. The result indicated that the amount of heat that could be produced in this way was virtually inexhaustible. So long as the boring continued, heat was produced. When the horses were pulling steadily against frictional resistance the rate of production of heat was steady. Rumford even calculated that the heat produced in this way was about the same as that produced by burning the food consumed by the horses – another remarkably prescient opinion that came very near to stating the principle of the conservation and transformation of energy. Rumford published this idea in a paper entitled 'Sources of the heat which is excited by friction'.

The following year, the young Humphry Davy published an account in which he claimed to have shown that two pieces of ice could be melted by rubbing them together in a vacuum. On the basis of this, he concluded, independently of Rumford, that heat was a form of movement – specifically, vibratory movement of the molecules of the material. This was a correct view but was based on faulty premises. A perusal of Davy's account shows that his claimed results were impossible. He was right, but for the wrong reasons.

In 1860 the Scottish mathematician and physicist James Clerk Maxwell (1831–79) decided to apply his mathematical talents to the problem of heat. Maxwell conceived of the molecules of a gas as being perfectly elastic and as moving randomly in all directions and at various speeds, bouncing off each other and off the walls of the container. Maxwell showed that the velocities of the particles followed a familiar statistical pattern – the bell-shaped curve. His work also indicated that a rise in temperature would increase the average velocity while a drop would decrease it. So it became apparent that Count Rumford and Humphry Davy were right and that the idea of heat as a fluid (caloric) could be finally abandoned.

The application of rays will gradually be taken out of the doctor's hands. He will write out a prescription, and we will go round to the radiologist's shop next door to the chemist's and ask for the prescribed treatment in his back parlour. The next man at the counter will be after an apparatus to radiate the buds of his rose bush and kill off insect's eggs without hurting the plants.

J. B. S. Haldane (1892–1964)

The n-ray blunder

When, around the end of the nineteenth century, Roentgen discovered X-rays and Plücker, Hittorf and Goldstein discovered cathode rays, the subject of rays was very much in the air. So it was no surprise when the French physicist M. R. Blondlot, announced in 1903 that he had discovered the n-ray.

This remarkable new ray could be detected when the emission from a normal X-ray tube was passed through a thin sheet of aluminium or black paper, so as to cut out visible light, and then allowed to fall on a small electric spark. The effect of the new ray was to cause the spark to brighten, so the spark was used as a detector. One remarkable thing about the n-ray was that it was polarized – restricted to a particular plane of vibration. Like visible light, it could be reflected and refracted. Extraordinarily, the n-ray had no effect on a photographic plate nor did it cause fluorescence on a standard fluorescent screen.

M. Blondlot hastened to report his discovery to the world and read a paper on it to the French Academy of Sciences in Paris. An account of this report was published in *Nature*. Naturally, the new ray aroused intense scientific interest. Every new form of radiation that had been detected to date had been found to be of the most fundamental scientific importance and its study had led to remarkable scientific advances.

Many physicists at once tried to reproduce M. Blondlot's findings. Strangely enough, letters describing their failure to do so soon began to appear in the scientific literature. No one had been able to show any trace of the remarkable and elusive n-ray. One sceptical scientist, the American physicist Robert Williams Wood (1868–1955), professor at Johns Hopkins University, decided to pay a visit to M. Blondlot's laboratory to see for himself. Once the experiment was set up, Wood was invited to agree that the ray was causing brightening of the spark and that when a hand was interposed the brightening diminished.

Regretfully, Professor Wood had to insist that he could not agree. As far as he was concerned, the spark was varying randomly in brightness and these variations bore no relation to the movement of his hand between the spark and the source of the n-rays. He was told that the reason for this was that his eyes were insufficiently sensitive. So Wood now turned the tables on M. Blondlot and suggested the latter tell him, by inspecting the spark, when Wood's hand was in the path of the rays. This was agreed.

Unfortunately, M. Blondlot got it wrong nearly every time. When Wood's hand was still the experimenter said the spark was changing in brightness, and when his hand was moving the report was that the spark was stable. Wood then observed that, when everything was left severely alone, the spark continued to fluctuate in brightness by an estimated 25 per cent. This is a common characteristic of sparks caused between electrical conductors, because the fact of sparking causes physical changes in the surfaces between which the sparks are occurring. Wood concluded that, as a sensitive detector of an alleged new ray, the electric spark was a non-starter.

Professor Wood was shown some photographs of the alleged brightening of the spark under the influence of the n-ray. He was then shown a photograph actually being made, but observed that, as well as the inherent fluctuation in the brightness of the spark, the plate used to make the picture was being shifted backwards and forwards every five seconds. In those days, of course, photographic plates were very insensitive and exposures had to be very long. This movement of the plate might well, he considered, give rise to a brightening of the picture by a cumulative favouring of one of the images, whether unconsciously or deliberately.

Professor Wood was then shown an alleged dispersion of the spectrum of the n-ray by means of a prism made of aluminium. Because the experiment was performed in the dark, the sceptical professor was able to perform a most ungentlemanly trick. In the course of the demonstration, he slyly removed the prism and was interested to note that this made no difference whatsoever to the appearance of the 'spectrum' on a screen. Having replaced the prism, he then requested permission to see whether turning the refracting edge to the right or to the left would alter the appearance of the spectrum. Failure to show any changes was attributed to incorrect placement of the prism as a result of fatigue on the part of M. Blondlot and his assistant.

Various other claimed phenomena were then demonstrated to Professor Wood but, sadly, although M. Blondlot insisted that the changes were obvious, the professor was unable to see them. Professor Wood left M. Blondlot's laboratory with the fixed conviction that the n-rays were a figment of the experimenter's imagination. Blondlot had laboured for a year to provide a mass of evidence to prove his case but had never, it seems, carried out any of the simple procedures that would have proved that the n-rays did not exist.

This sad case is quite instructive. It shows us how powerfully the desire for scientific fame can interfere with the detachment and disinterestedness required of the true scientist. Blondlot had held firmly on to every scrap of evidence that seemed to support the hypothesis that a new ray existed, but had studiously ignored any findings that went against it. Worse than this, he had clearly been driven to modify his experimental procedure in such a way as to give rise to supportive results.

There are only two possibilities here: either Blondlot was monstrously fooling himself or he was a conscious fraud. Unless he was also a totally unrealistic optimist who thought he could get away with it, the latter alternative seems unlikely.

There is an ethereal medium pervading all bodies. The parts of this medium are capable of being set in motion by electric currents and magnets.
James Clerk Maxwell (1831–79), the greatest theoretical physicist of his time, whose equations established the theoretical foundations of electromagnetism

---

The importance of ether

Nowadays, the term 'ether' brings to mind a strong-smelling, volatile and rather old-fashioned anaesthetic liquid. Present-day chemists are, of course, still interested in ethers, but to the physicists of the nineteenth century the ether was a concept of central importance. To avoid confusion with chemistry, the physical ether was often spelled 'aether' or was known as the luminiferous ether or aether. To the physicists the ether was very real, even more real than matter. It was the universal all-pervading substance, filling all space and indeed all matter, that was believed to act as the medium for transmission of light and other electromagnetic waves.

To the physicists of the time, it was inconceivable that any kind of waves could be propagated except by way of the medium in which the waves were occurring. This was only common sense. How could you possibly have waves at sea without the sea? Waves in water and in air had long been studied and were well understood. Waves could exist only in an elastic medium such as a liquid or a gas. Sound waves and other vibrations could also be transmitted through solids. So it was obvious that light, being a wave phenomenon, also had to have a medium to vibrate. This was the basis on which the ether was postulated. Note that the ether was not discovered; it was simply necessary, so it had to have a name.

Since the ether had never been seen, felt, or otherwise detected it was known to be transparent, weightless, frictionless, and devoid of chemical or physical properties. It did, however, permeate all space and matter, and, as well as being the medium of light and heat, was the vehicle of gravitation. This made it a remarkable entity, quite different from anything else known to science. There were a few heretics, some of them quite distinguished, who doubted the existence of the ether. These difficult people pointed out that the properties of the ether could only be explained by attributing contradictory facts to it. However, they were laughed out of court by the majority of scientists who really understood the matter.

One of the great physicists of all time was James Clerk Maxwell (1831–79). It was Maxwell who showed that oscillating electric charges could generate electromagnetic waves that were propagated at the speed of light and who conclusively demonstrated that light was an electromagnetic phenomenon. These waves consist of alternating electric and magnetic fields at right angles to each other. Maxwell's famous equations described the wave propagation mathematically for all time. This was, of course, the basis of radio and TV. Maxwell never doubted the existence of the ether, and was apt to wax lyrical about it. He would describe its vast expanse, how it was everywhere in the universe, passing right through the planets in their orbits 'as the water of the sea passes through the meshes of a net when it is towed along by a boat'. Had he lived a little longer, Maxwell would have regretted this poetical expression of his, which was immortalized in the ninth edition of the *Encyclopaedia Britannica*.

Maxwell was, of course, deeply interested in finding out more about an entity as fundamental to all science as the ether. Accustomed to think big, he came up with a remarkable experiment to discover how movement through the ether affected the speed of light. He pointed out that light moving through the ether in the same direction as the movement of the earth in its orbit round the sun must differ in speed from light moving at right angles to that direction, and this difference should be measurable. Discovering this difference would tell us quite a lot about the 'drag' of the ether.

Eight years after Maxwell died, his experiment was tried. Two physicists, Albert Michelson (1852–1931) and Edward Morley (1838–1923), using an instrument called an interferometer, checked the speed of light in these two directions. The instrument was more than amply sensitive to deal with ratios as large as those of the speed of the earth and the speed of light. As Maxwell had indicated, since we knew the speed of the earth in its orbit, the speed of light in the two directions should differ by about one part in 100 million. The interferometer could detect differences to an accuracy of one part in four billion. To the astonishment of Michelson and Morley and to the disbelief of the rest of the world of physics, no difference in the speed of the light was found. However often the experiment was done, the result was the same. No difference.

It was obvious that Michelson and Morley had done something wrong. Other physicists tried the experiment, only to get the same negative result. Then someone came up with the explanation. Of course, the ether was being dragged along with the earth and was thus stationary relative to the interferometer. Problem solved. For nearly 40 years, until about the mid-1920s, scientists went on trying to detect the ether. But by then a few important things had happened. The failure of the Michelson–Morley experiment had set up an extraordinary train of thought in the mind of a young German-Swiss patent office clerk called Albert Einstein. The result was the special theory of relativity. This asserted that the speed of light was constant under all conditions and accounted for the negative result of the experiment without having to make any reference to the now rapidly fading ether. Moreover, a young German physicist called Heinrich Hertz (1857–94) had reformulated Maxwell's theory to make it clear that electromagnetic radiation did not require any sort of medium for its propagation. By 1930 younger physicists would smile in a supercilious fashion at any reference to the ether.

All scientists now agree that, in the words of the American homespun philosopher: 'They ain't no sech critter.'

I utterly reject the atomic theory of Dalton.

Sir Humphry Davy (1778–1829), the outstanding chemist and scientific propagandist of his time, who ignored the incontrovertible facts on which Dalton's theory was based and brought out objection after objection

---

Atomic notions

From the earliest times, thoughtful people have considered it probable that there must be a limit to the extent to which anything could be repeatedly divided – that, in other words, there must be a smallest size for everything. Thought about this led to the idea of the atom (from the Greek *atomos* meaning 'not able to be cut'). The idea of very small particles from which everything is made was mooted at least 2500 years ago, which is rather remarkable considering that it was not until the end of the nineteenth century that any real facts about the nature of these particles became available.

The first well-known account of a theory of atoms was proposed by the Greek scientific thinker or 'natural philosopher' Democritus (*c.*470–380 BC). Actually, there is evidence that Democritus cribbed this idea from his teacher Leucippus of Miletus who flourished in the fifth century BC. Hardly anyone has ever heard of Leucippus while Democritus is quite well known, so naturally, the latter gets the credit for this important idea. Democritus taught that all matter was made up of particles so small that nothing smaller could be conceived. These particles, which he called atoms, were indestructible and, of course, could not be cut. They were solid, hard and incompressible, and each type of material was made up of large numbers of individual atoms. A pure material would consist of only one type of atom in huge numbers. More complex materials contained a range of types of atoms. Atoms were of different shape and it was this that gave them their properties. For instance, white material had atoms with smooth white surfaces. A sour taste was caused by needle-sharp atoms. The human soul was made of atoms that were smaller and finer than any others.

Democritus taught that atoms could combine together, in accordance with strict laws of nature, to form different substances. Although fanciful in some of its details, the atomic theory of Democritus (or his boss) was ingenious and was capable of providing some kind of an explanation for a number of facts. It was also a great advance on the rather feeble magical thinking of other philosophers.

So, in that respect, it was, for the time, quite good science; but as a fully adequate explanation, it had to be scored pretty low.

Most of the Greek philosophers were, unfortunately, convinced that natural science was a rather low-grade activity and distinctly *infra dignitatem*, as the Romans would have put it. It was fine for slaves and mechanics and other blue-collar workers who didn't mind getting their hands dirty, but it was not quite the thing for gentlemen. So for a couple of centuries natural science languished while Greek philosophy, under the influence of Socrates, turned to moral and metaphysical questions.

The atomic theory of Democritus was not wholly forgotten, however. The Greek philosopher Epicurus (341–270 BC) thought it was rather good and even found it useful. Epicurus, like many people of today, took a mechanistic view of the universe. Greek superstition and belief in magic and in the power of the Gods offended him and he strongly advocated Democritus' atomic theory to try to counter this superstition. If, as he maintained, the entire universe consisted only of atoms and nothing else, then even the gods were made of atoms and were subject to exactly the same scientific laws as were humans. So there was no need to worry.

Another important advocate of the general idea of atoms was the Roman poet and philosopher Titus Lucretius (*c.*99–55 BC). He put the notion in rather a fine way in his great scientific poem *De rerum natura* (On the Nature of Things). To illustrate the idea of atoms, Lucretius described how he stood on a high hill watching an army on a plain so far below him that it resembled a single massive solid body glittering in the sun. Although it looked solid, he was, of course, aware that it was made up of an enormous number of individual parts. Lucretius' poem was almost lost and was unknown throughout the Middle Ages. But a single manuscript copy survived and, soon after the invention of printing, the poem was published in full in

1417 and became widely popular. The idea of atoms was thus available to the thinking of many educated people after the Renaissance.

Another reason for the preservation of this idea was, ironically, the powerful opposition of the immensely influential Greek philosopher and scientist Aristotle (384–322 BC). It has to be remembered that very few of the Greeks were scientists in the sense we use the term today. They did not make observations or carry out experiments. In their view, only pure thought was truly worthy. So, to them, science consisted of dreaming up intellectual theories that would explain nature. Aristotle was very good at this kind of thing. Furthermore, he hated Democritus' idea of atoms.

For nearly two thousand years after his death Aristotle was by far the most widely read of the philosophers and his views carried great weight. He was also remarkably popular because, for many, he was the only source of information on such matters as physiology, logic, ethics, the acquisition of wealth, politics and psychology. Aristotle's attack on Democritus' atomic theory was not based on any real scientific grounds but was grounded in pure prejudice and speculative philosophy. This, he believed, was how knowledge was to be obtained. Aristotle also had the backing of the Church and anyone who rejected Aristotle was liable to be in serious trouble. In particular, the Roman Catholic theologians decided that the atomic ideas of Democritus were not only materialistic; they were frankly atheistic.

So atoms had a bad press for many centuries until the growth of real science began to look again at the question. In the seventeenth century the popularity of *De rerum natura* and the growth of scientific, and especially chemical, knowledge helped to maintain a debate between what was then considered the orthodoxy of Aristotle and the revival of the ideas of atomism. At the beginning of the nineteenth century the idea of atoms became one of practical importance to the chemists. The English schoolteacher John Dalton (1766–1844), who was interested in all branches of science, proposed that 'the ultimate particles of all homogeneous bodies are perfectly alike in weight, figure *et cetera*'. By 'homogeneous bodies' Dalton did not mean only elements, but included compounds such as water. So, in this, he was not distinguishing atoms from molecules. He was, however, aware that the atoms of an element were, in a chemical context, indivisible and that they remained unchanged while undergoing chemical reactions. He went further. Taking hydrogen as the lightest element, he was able to work out the relative weights of the atoms of other elements, such as oxygen, nitrogen and sulphur.

Dalton's ideas advanced scientific thinking considerably and, because he used his ideas of the atom to explain other phenomena, he can quite realistically be considered the father of modern atomic theory. But were Dalton's ideas of the atom adequate? Regrettably, no. There were too many unanswered questions and, as science advanced, these increased in number. One question that arose from scientific advance was how to account for the extraordinary fact that certain elements, when purified, were not only warmer than their surroundings but were giving off energy. Two years before the end of the nineteenth century, Marie and Pierre Curie had discovered the radioactive elements polonium and radium. Up to this point, everyone had accepted the idea that atoms were the ultimate small entities. By definition they were indivisible. That's what they were called. But here were atoms in which something very odd was going on and it began to look likely that these ones, at least, were rather more complicated than people had imagined.

At about the same time as the discovery of radium another remarkable new fact appeared. In 1897, the English physicist Joseph John Thomson (1856–1940), known to his colleagues as 'J. J.' had been working with cathode ray tubes of the kind devised by William Crookes. Thomson knew that the rays, which produced a fine spot on a fluorescent screen on the end of such a tube, could be deflected by magnets and by an electric charge on plates within the tube. Changing the polarity of the charge showed that the beam was attracted by a positive charge and repelled by a negative charge. So, because 'like' charges were known to repel each other, the beam itself had to carry a negative charge. Thomson assumed that the beam consisted of a stream of negatively charged particles, or 'corpuscles', as he called them. Measuring the deflection of the spot allowed him to quantify the ratio of the charge to the mass of one of these particles. This was shown to be less than one-thousandth of the mass of a hydrogen atom. In other words, particles existed that were smaller than an atom. Shock horror!

For a time, no one believed J. J. The indivisibility of the atom had been such a firmly entrenched dogma throughout the whole of the nineteenth century, that when, at a meeting at the Royal Institution in 1897, Thomson announced the discovery of the electron, a distinguished physicist told him afterwards that he thought Thomson had been pulling their legs.

As soon as it became known that the atom contained particles smaller than itself, speculation arose as to its structure. J. J., of course, speculated on this question. His tentative suggestion was that the atom consisted of a hard ball of positive electricity with electrons stuck on to it, or embedded in it, like currants in a bun. Was this an adequate description of the atom? Unfortunately not. It raised even more questions than it answered.

Working briefly under J. J Thomson at Cambridge was the New Zealand student Ernest Rutherford (1871–1937). Rutherford was a high-flier who at 27 became professor of physics at McGill University and by the age of 37 was a Nobel Prizewinner. Interested in the then new field of radioactivity, he knew that the rays given off by radioactive materials were of different kinds. He considered this important. So, quite arbitrarily, he called the positively charged rays 'alpha rays' and the negatively charged ones 'beta rays'. Later, between 1906 and 1909, assisted by the young Hans Wilhelm Geiger at Manchester University, he was able to prove that alpha rays were streams of helium atoms minus their electrons. These particles had a double positive charge.

In 1910, Geiger and another of Rutherford's students fired streams of alpha particles at a very thin sheet of gold foil. Most of them passed through and were registered on the photographic plate behind. But a very small number – about one in 20,000 – actually bounced back. This was an astonishing finding. 'It was almost as incredible,' said Rutherford, 'as if you fired a fifteen-inch shell at a sheet of tissue paper and it came back and hit you.'

These observations led Rutherford to two important conclusions: that atoms were mostly empty space and that they had a positively charged centre, somewhat similar to the alpha particles. Because like charges (two positive or two negative charges) repel, alpha particles that happened to strike a positively charged atom centre were forced back. Rutherford now felt able to propose a new model for the structure of the atom, and he did so in 1911. It consisted, he suggested, of a positively charged central part, the nucleus, occupying only a tiny proportion of the whole volume of the atom, surrounded by a large space containing negatively charged electrons. He was able to work out that the mass of the positively charged particle of the nucleus, which he called a proton, was more than 1800 times that of an electron. So almost all the mass of the atom resided in the nucleus. Because electrons were of opposite charge to the nucleus and would be attracted to it, they had to have energy of their own in the form of rapid movement around the nucleus. The analogy with the solar system was irresistible and the electrons came to be described as 'planetary electrons'. Their angular velocity in their orbits round the nucleus, he suggested, provided just the required degree of centrifugal force to balance the attraction of the nucleus.

For each proton in the nucleus there was one orbital electron, so the atom remained electrically neutral. Hydrogen had one proton in the nucleus and one planetary electron. Helium had two protons and two electrons. Lithium had three protons and three electrons, Chlorine had 35 protons and 35 electrons, and so on. So it was the number of electrons in the atom that determined the chemical properties, and the number of electrons was ultimately determined by the number of protons in the nucleus.

Rutherford knew that the helium nucleus had twice the mass it would have if the two protons were all it contained. At first, he suggested that it might contain four protons, two of which were neutralized by two electrons. But there were various objections to this explanation and he had reason to believe that an uncharged (neutral) particle of the same mass as a proton actually existed. In the 1920s Rutherford and his assistant James Chadwick (1891–1974) spent a long time looking for such a particle – which could be called a 'neutron' – but failed. Uncharged particles were very difficult to detect simply because they *were* uncharged. They could not be directly detected by any of the electrostatic methods then in use. Chadwick eventually proved the existence of the neutron in 1934 and won a Nobel Prize and a knighthood for himself.

Rutherford's concept of the atom immediately provided explanations for many well-known chemical and other phenomena and was one of the most germinal and fruitful hypotheses in the entire history of science. It was a great leap forward and advanced physical science and chemistry enormously. But was it true?

For a time it seemed the complete answer. On the basis of this model, scientists were soon able to make the following important statements. Hydrogen is the only atom with no neutron in the nucleus. Helium has two protons and two neutrons. Lithium three protons and three neutrons. The atomic number can be taken to be the number of protons and this rises by one with each different heavier element until we reach uranium with 92 protons – the heaviest of the naturally occurring elements. The number of neutrons, however, is not always the same as the number of protons. Many of the heavier atoms have more neutrons than protons and many atoms with the same number of protons (i.e. of the same element) have different numbers of neutrons. Most samples of uranium, for instance, have a mass equal to 238 protons because the nuclei contain 92 protons and 146 neutrons. Some samples – the kind used in the early atom bombs – have a mass of 235 with 92 protons and only 143 neutrons.

The chemical properties of an atom depend on how it links with other atoms by way of its electrons. So these properties depend on the number of electrons and, consequently, on the number of protons in the nucleus. The chemical properties are quite unaffected by the number of neutrons. Atoms with the same number of protons but different numbers of neutrons are called isotopes – literally 'equally placed' (in the periodic table). The physical properties, however, depend also on the number of neutrons. Very heavy atoms with many neutrons are often unstable and can break down, for example by giving off alpha particles (two protons and two neutrons) from the nucleus. The loss of two protons, of course means a loss of two electrons and consequently a complete change to a different element with different chemical properties. This is called transmutation. Elements that undergo spontaneous changes of this kind are said to be radioactive. Some isotopes can also be radioactive.

Rutherford was awarded the Nobel Prize in 1908 for his work on radioactivity and was knighted in 1914. He became a professor of physics at Cambridge in 1919 and succeeded J. J. Thomson as Director of the Cavendish Laboratory. In 1921 he received the Order of Merit, he was President of the Royal Society from 1915 to 1930 and was raised to a peerage, as Baron Rutherford of Nelson, in 1931. But was his model of the atom correct?

Unfortunately, far from being a complete account of the nature of the atom, consistent with all other knowledge, Rutherford's model was far from accurate and it was soon apparent that it would not do at all. There were certain important facts which it simply could not explain. If we view the atom as a kind of miniature solar system, we immediately run into a major problem. Orbiting electrons must, according to classical theory, emit energy in the form of electromagnetic radiation. All moving electric charges emit radiation. This is how radio and TV work. But if electrons gave off energy in the form of radiation they would have less to keep them spinning round in their orbits and would spiral down into the nucleus. Their kinetic energy (the energy of movement) would decline and they would fall.

Remember that electrons are negative, the atomic nucleus is positive, and that unlike charges attract each other. In Rutherford's model (which seemed so convincing that it came to be called the classical theory) the only thing that kept the electrons from being attracted by the positive charge on the nuclear protons and plunging into the nucleus was their kinetic energy. This, he explained, supplied a kind of centrifugal force. In fact, although spinning electrons do give off energy, they don't spiral down into the nucleus. Atoms certainly emit electromagnetic radiation but they do so at frequencies specific to the type of atom (see below) and they go on doing so. This was too big a camel to swallow. Rutherford's atom just wasn't good enough.

Another problem troubling the physicists was the demonstrable fact that you could do an experiment to prove that light was a wave phenomenon. You could, for instance, show the expected interference between two sets of light waves just as you can show interference between sound waves and even sea waves. But you can also do an experiment to prove that light consists of particles. Rutherford's model of the atom has nothing to say to us on this seemingly fundamental difficulty.

Other things were worrying the physicists. Principal among these was the very odd fact about the way hot bodies gave off energy. When you heat a bit of iron it gets red, then orange, yellow, green, blue and violet. White heat is just a mixture of all these colours. The colour change through the spectrum from red to violet is simply a matter of the wavelength of the energy given off. Red means long wavelengths, yellow shorter, blue shorter still and violet shortest of all in the visible spectrum. Now, the shorter the wavelength, the more energetic the wave. So, according to classical theory, radiation at the violet end of the spectrum should have a lot of energy and radiation beyond that should continue to rise steeply.

The visible spectrum is only a tiny part of the whole electromagnetic spectrum, which extends a long way on either side of visible light. In terms of wavelength, there is a great region of infrared radiation, of increasing wavelength, below the red; and above the violet, there is a great region of ultraviolet radiation of decreasing wavelength, to say nothing of the X-rays and gamma rays beyond that. By classical theory, then, the energy of radiation should, somewhere in the ultraviolet, reach catastrophic levels. As a consequence, this idea became known as the 'ultraviolet catastrophe' and no one had the least idea why it didn't happen. In fact, simple measurements showed that, within the visible spectrum, as the wavelength decreased, the energy, after rising at first, began to fall again. The ultraviolet catastrophe remained a painful puzzle for years.

In October 1900, the German physicist Max Planck (1858–1947) took a walk in the Grunewald woods outside Berlin. Like all his physicist colleagues, Planck had long been worrying about these and other facts that did not make sense on the basis of classical Newtonian physics. He was thinking hard how to balance an important mathematical equation that had stubbornly refused to balance. When he returned from his walk, he modestly wrote down: 'Today I made a discovery as important as Newton's discovery of gravitation.'

Under Newtonian physics the emission of energy – light, heat and other forms of radiation – was assumed to be continuous. There was no reason to think otherwise. What Planck postulated was different. Energy, he decided, was given off in a series of very small separate packets, which he called 'quanta'. This was a crazy idea, but it fitted nicely with certain incontrovertible facts that could not be explained otherwise. Planck knew that the amount of energy carried in a particle was directly related to the frequency (number of cycles per second) of the wave. Remember that frequency and wavelength are completely bound up in each other. They are reciprocally related: as one increases, the other decreases. If you double the number of wiggles that fit into a given period of time, each of the new wiggles must be half the length of the previous ones.

Planck's Grunewald insight was that the energy was equal to the frequency multiplied by a very small number h – a constant that is now universally known as Planck's constant. That tiny number was to prove important enough not only to alter fundamentally the ideas about the nature of the atom, but also to turn classical physics upside-down.

Planck's idea of quanta sounded like nonsense at first, but it did provide a way of answering the riddle of the ultraviolet catastrophe. At high frequencies (or short wavelengths, such as those in the ultraviolet) a great deal of energy would be needed to emit a quantum. Only a few of the energy emitters of atoms – the electrons – would be energetic enough to supply this amount of energy. At low frequencies, there are masses of electrons with enough energy to emit quanta of low energy. Somewhere in between would be a peak. The mathematics fitted nicely, but the whole idea worked only on the ridiculous basis that energy was given off in packets. As everyone knew, electromagnetic radiation was a wave.

Einstein, investigating how light falling on certain materials, such as selenium, would cause a small electric current to flow (the photoelectric effect), had shown that a certain precise minimum amount of light energy was needed to knock an electron off an atom. He also showed that the kinetic (movement) energy of the electron flying off was equal to the energy of the knocking-off photon minus the energy needed to do the knocking-off. Planck's new idea led to the theory that, in an atom, each electron is in a certain energy state and can move to a higher energy state only by absorbing a precise quantum of energy. As it does so, it jumps instantaneously to the higher energy level. There is no question of a particle acquiring a smoothly varying amount of energy. It can only make a quantum leap – a tiny but precise change in its energy.

This idea nicely dealt with the problem of how electrons could give off energy and still stay *in situ*. They only gave off energy that they had received from outside the atom. The ideas answered a lot of other questions, but the mare's nest of new problems it uncovered proved to be unprecedented in the whole history of science.

Some of the consequences of quantum theory are virtually un-believable. Take the change in energy level of an electron. (If you want to continue to think of an atom as being like a solar system with the sun as the nucleus and the electrons as the planets, you can safely think of different energy levels as being different 'orbits'.) Bohr proposed that electrons can move only in certain permitted orbits and while in these orbits do not emit radiation. The energy of an electron in a particular orbit is definite and consists of two parts – its potential energy by virtue of its distance from the nucleus, and its kinetic energy from its movement. Each permitted orbit, therefore, is associated with a particular level of energy. An electron, he suggested, can move suddenly from an orbit of higher energy to one of lower energy. When it does so, the energy difference is emitted as a quantum of electromagnetic radiation, such as light, for instance, of a particular frequency.

Every element emits its own characteristic light wavelength when heated. The sodium in common salt gives a yellow colour, for instance, because when the excited electrons in the sodium atom return to their non-excited level, they give off energy at precisely the frequency of blue light. Heating supplies energy and raises electrons to a higher energy level. When they fall back they emit light of a precise wavelength that is determined by the structure of the atom. Checking the wavelength of the light allows us to identify the atom. This is the basis of spectroscopy – the technique that had allowed scientists for many years to tell what distant stars are made of. The new quantum theory image of the atom provided an explanation of this phenomenon.

Gradually, Planck's idea took hold and, with it, Bohr's model of the atom. Once scientists had grasped its principles, the Bohr atom took the scientific world by storm. The physicists immediately got to work designing experiments to prove that it was right. Success quickly followed success. The idea of the permitted orbits with their ladder of energies was proved by James Franck (1882–1964) and Gustav Hertz (1887–1975) – the nephew of Heinrich – and won them the 1925 Nobel Prize. Using gaseous mercury, which they bombarded with electrons, they showed that the energy was absorbed by the gas in discrete amounts (quanta) of 4.9 electron-volts. This

[handwritten margin note: yellow]

caused the mercury to get excited and then to return to its original state after giving off a photon of light of precise wavelength.

The result of this experiment was, of course, a great encouragement both to Bohr and to Max Planck and gave them additional stimulus to go on developing the idea of the atom and quantum theory respectively.

So was the Bohr atom, with its basis in the newly established quantum physics, the last word? Was it the complete answer? Regrettably, no. Bohr's model still had many shortcomings, and, in spite of its power, was destined to be swept aside a mere 12 years after it was first announced. The Bohr atom model could not account for the spectral lines of atoms with more than one electron – that is, atoms heavier than hydrogen. Furthermore, it did nothing to account for the extraordinary wave/particle problem – the fact that there was clear experimental evidence that light behaved both as a wave and as a particle.

It is time to introduce an aristocrat – a prince, no less, who later became a duke. Louis-Victor Pierre Raymond, Duc de Broglie (1892–1987) (pronounced 'broy') was a nobleman whose great-great-grandfather had the distinction of having been guillotined during the French Revolution. De Broglie was expected to enter the diplomatic service or the army and was sent to the Sorbonne to read history. But he had already become interested in science because of the work his elder brother was doing on X-ray spectroscopy in his private laboratory, and wasted much of his study time immersed in science books. Nevertheless, he got his history degree. During World War I, while still a mere prince, he was posted to the Eiffel Tower radio station and this got him even more interested in science. So after the war, he returned to the Sorbonne, where, in 1924, he took a doctorate in physics. He very nearly didn't because his doctorate thesis was so far ahead of his professors that some of them thought it was rubbish. Fortunately, they sent a copy to Einstein who said, in effect: 'This is good stuff.' So the prince became a doctor and a totally new concept was presented to the world.

By now, Einstein's celebrated equation $E=mc^2$ was well known. Energy equals mass multiplied by the speed of light multiplied by the speed of light. Everything that has mass has energy. Particles, such as electrons, have mass so they also have energy. Planck had pointed out that energy was equal to frequency multiplied by a very small number called Planck's constant: $E=hv$, where h is Planck's constant and v is the frequency. Everyone knew this, too.

So now it was de Broglie's turn. All matter, he suggested, must display wave-like properties – must, indeed, act like waves. To him it seemed obvious. Energy implies frequency; frequency implies waves; therefore particles must behave as waves. How did he explain this incredible suggestion? Simple. Einstein's equation and Planck's equation are not two separate statements; they are interrelated. If you know the mass of something, you multiply it by the speed of light squared and you get the energy. And if you divide this energy by Planck's constant you get the frequency. So every particle has a definite frequency or rate of pulsation associated with it.

If you consider a wave, it can be thought of as a simple up and down motion, like that of a cork on a pond when a stone is thrown in. But it can also be thought of as the outward propagation of the ripples from around the point at which the stone dropped. De Broglie incorporated both ideas. A particle at rest has a local up and down vibration and also a wave that is propagated outward to infinity. Movement of a particle, at speeds much less than the speed of propagation of the wave, can be interpreted as the movement of the resultant wave formed by the interference of many waves whose frequency had to differ slightly in relativistic terms. Matter-waves eluded experimental demonstration for a time, but in 1927 they were actually detected.

So the Bohr model of the atom had to be superseded by a new model, proposed by the Austrian physicist Erwin Schrödinger (1887–1961) based on his new discipline, wave mechanics. Schrödinger's atom incorporates Louis de Broglie's concept of the electron as having wave properties. Electrons can be in any orbit around which an exact number of wavelengths can occur, setting up what is called a 'standing wave' like the sound waves in an organ pipe. As there was no accelerating charge, there was no radiation. 'Permissible' orbits were determined by the need for the exact number of wavelengths to be present. Other conceivable orbits would involve more or less than a whole number of waves and so would not occur. Schrödinger's model, published in 1926, offered a more rigorous and mathematically sound account of the atom, than that of Bohr. All three men – Bohr, de Broglie and Schrödinger – were awarded well-deserved Nobel Prizes.

So, has the Schrödinger model of the atom reached the end of the line? Have we finally arrived at an unalterable picture of the atom? Happily not. That's not what science is like.

> Atomic energy might be as good as our present day explosives, but it is unlikely to produce anything very much more dangerous.
>
> *Winston Churchill (1874–1965), speaking in 1939*

The cold fusion affair

When the power locked up in fissionable matter was so brutally demonstrated over Hiroshima in 1945, the world was persuaded, mainly by journalistic hype, that its energy problems were at an end. Endless quantities of cheap power were there for the taking. The reality proved somewhat different. Power stations based on nuclear reactors turned out not only to be at least as expensive as conventional fuel power generators but also to have some very nasty characteristics. All of them produced highly radioactive and dangerous long-life nuclear waste and some of them even bred larger quantities of radioactive isotopes than they consumed. Breeder reactors produce plutonium, one of the most toxic and dangerous substances known. Plutonium-239 has a half-life of 24,000 years: this means that a given quantity of plutonium will decay by half by emitting radiation in this time.

But nuclear fission is not the only way to produce power from the atom. As was soon alarmingly demonstrated in test explosions of the hydrogen bomb, vast quantities of power could be released when, instead of splitting the nuclei of heavy atoms into lighter elements, light atom nuclei, such as those of hydrogen isotopes, are persuaded to combine to form nuclei of other elements of heavier atomic weight. In both cases, a small quantity of matter is converted directly into energy. Since $E=mc^2$ and c (the speed of light) is an enormous figure even before it is squared (multiplied by itself), the energy produced by the conversion of a small amount of matter is very great. But there is a most important difference. Fusion reactions are more or less 'clean': these reactions do not usually involve radioactivity. This makes them ecologically and politically much more attractive as a basis for power generation than fission reactions.

By exploding hydrogen bombs, scientists showed in the 1950s that fusion reactions were possible and that they produced fabulous amounts of the energy. But they also began to demonstrate the difficulties involved. To achieve a fusion reaction, it is necessary to force the two reacting nuclei together strongly enough to get them to link up. Since they are both positively charged and since like charges repel each other this is difficult. The enormous repulsive force between them can only be overcome if they are brought so close together that they get within the range of the 'strong' nuclear force. Unfortunately, the strong nuclear force has a very short range – only about the diameter of a hydrogen atom. Once within this range, fusion will occur.

The most obvious way to achieve this degree of intimacy is to up the heat. Heat is movement. If the nuclei can be made to move at very high speeds – that is, to have very high kinetic energies – they can get close enough to fuse. Such energies imply temperatures of the order of 10 million degrees Celsius – the kinds of temperatures occurring in the sun and other stars. Indeed, it was only when fusion reactions were discovered that scientists were able to explain why the sun had not burned itself out long ago.

At present, the only easy way to achieve temperatures of this order is to explode a nuclear bomb and this is how the hydrogen bomb was made – a layer of 'heavy hydrogen' (deuterium) surrounding a normal plutonium bomb. To call this a 'clean' bomb is to stretch the truth a bit. The energy from the fusion part is clean, but this is obtained only by a very dirty fission reaction. But for years now, scientists have been working on machines that can contain such temperatures. These are necessarily large and extremely expensive, and, although fusion reactions have certainly been produced, none of them has worked properly yet in the sense of producing a continuous, self-sustaining and usable net increase in energy production.

There is, of course, no material from which a container capable of sustaining these temperature can be made. The idea behind these machines is to produce an immensely powerful magnetic field to contain the fusionable material. At temperatures above about 10,000 degrees, the electrons get stripped off atoms leaving the positively charged nuclei and free electrons. This ionized state of matter is called a plasma. Plasmas can conduct electricity and can be confined by a dough-ring shaped (toroidal) magnetic field. The stronger the field, the more tightly compressed and hotter become the individual charged particles of the plasma. Various different designs are possible and several of these have been tried at enormous expense. These include the neutral-beam-heated 2X-IIB minimum-B mirror machine at the Lawrence Livermore National Laboratory in California, the Tokamak Fusion Test Reactor (TFTR) at Princeton, New Jersey, and the Joint European Torus (JET) in Culham, England. In the USA alone more than \$6 billion has been spent on these machines.

In the light of all this scientific background, it will not be hard to imagine the kind of reaction produced in the scientific community by a claim by two chemistry professors that they had achieved nuclear fusion at normal room temperatures. In March 1989 Martin Fleischmann of Southampton University and B. Stanley Pons of the University of Utah revealed what they claimed to be a simple method of achieving cold nuclear fusion. First reports of major scientific discoveries are normally made in scientific journals and, before publication, are sent to experts in the field for comment. This is called 'peer review'. Unless the experts agree that they are likely to be genuine they are nearly always denied publication by respectable journals. Fleischmann and Pons bypassed this process and announced their claim at a press conference at the University of Utah. Inevitably, it was hailed by many journalists as the greatest scientific breakthrough for decades. This was finally the definitive source of unlimited cheap energy.

Not only that, but the new power unit was small and portable. It could fit in a car or a private house and would clearly revolutionize energy production. The implications – economic, ecological, social and political were hard to exaggerate. Public interest was immense and, for the first time, many politicians, business people and non-scientific journalists tried desperately to grapple with the complexities of nuclear physics. To understand the claims of the two scientists, they had to grasp, for instance, that two nuclei of atoms of deuterium (heavy hydrogen or hydrogen-2) could fuse to form one nucleus of even heavier hydrogen, tritium (hydrogen-3), and a single proton (a normal hydrogen nucleus), with the release of 4 million electron-volts of energy.

The method adopted by the two scientists could be done on an ordinary laboratory bench and involved the passage of an electric current through a solution of lithium in heavy water (deuterium oxide) using electrodes of platinum and palladium. The platinum anode (positive electrode) was in the form of a coil of wire around a palladium cathode (negative electrode). The current carried the deuterium nuclei into the crystal lattice of the palladium, where they were squeezed together so tightly as to become fused. The energy produced was released in the form of heat, and more energy was produced than was supplied in the current of electricity passing through the equipment.

Heavy hydrogen is chemically identical to normal hydrogen in that there is only one proton in the nucleus and hence only one electron in the orbit. It is the number of electrons that determines the chemical properties of an atom. But heavy hydrogen differs physically in that it also has one neutron (deuterium) or two neutrons (tritium) in the nucleus. A neutron is a particle of about the same mass as a proton, but with no electric charge. When two deuterium nuclei fuse, tritium may be formed and a proton emitted, or helium–3 may be formed and a neutron emitted. Helium is the next element above hydrogen and has two protons and usually two neutrons in its nucleus. It is named after the sun where it is found in plenty as a result of the fusion reactions going on there.

Fleischmann and Pons insisted that their experiment had resulted in the net production of energy as heat and in the production of neutrons, tritium and helium. Needless to say, the scientific community was highly sceptical of the claim. On the face of it, the whole thing was too simple and much too improbable. Even so, several hundred physics and chemistry laboratories immediately tried to reproduce the experiment and, at first, there were some confirmations. But only a few. Most of the laboratries reported the not-unexpected lack of success. Laboratories that turned the thumb down included the Harwell Nuclear Research Centre at Didcot, the Lawrence Livermore National Laboratory in California, the Los Alamos National Laboratory and the Massachusetts Institute of Technology.

Soon all kinds of objections and criticisms arose. The two chemists claimed to have produced 4 watts of heat for every watt of input power. But the number of neutrons they claimed to have counted was less than a hundred per second. Heat output by the watt should have been associated with many billions of neutrons per second – a radiation that would have done Fleischmann and Pons quite a lot of obvious harm. Yet, there they were, seemingly as good as new. Another criticism was that the equipment Fleischmann and Pons used to count neutron output – a rather crude device called a BF3 – was very sensitive to heat. If brought near a heat source, it would start to register a higher neutron count solely as a result of the heat. This anomaly was responsible for the results of one of the laboratories that had claimed to have confirmed Fleischmann and Pons's findings.

Less than a week after the press conference, Fleischmann gave a lecture at Harwell – the British Atomic Energy Establishment. There was one significant event that caused the scientists to glance meaningfully at each other. One of them asked Fleischmann if they had done control experiments with ordinary water instead of heavy water. This was an elementary step. If such a control trial had produced zero results compared with the definitive experiment, this would have strengthened the credibility of the whole claim. But if the control had produced exactly the same result, it would cast strong doubt on their claim that heavy water was essential.

Instead of saying yes or no, Fleischmann said: 'I'm not prepared to answer.' This response caused a stir and soon afterwards someone asked the same question of Pons. He had a slightly better answer. A control experiment with ordinary water, he said, would not necessarily be a good baseline. Again, eyebrows were raised.

In spite of all the scepticism, some laboratories got very excited. Scientists at the University of Moscow reported that the neutron levels produced were five times as high as background. A laboratory in Texas claimed to have found tritium and excess heat. Several others claimed success and, for a time, the world scientific community was divided into two camps. Numerous letters appeared in the correspondence columns of serious scientific journals arguing for and against the claim. The consensus appeared to be against. Most laboratories either found no effect or were convinced that the results they did find were due to non-nuclear reactions. Even so, there was so much support for the two chemists that a National Cold Fusion Institute was set up with a grant from the Utah legislature of $5 million.

In November 1989 a large advisory panel of the US Department of Energy reported that the results found did not provide convincing evidence that the heat produced was due to nuclear reactions. One thing was certain: no one had managed to produce useful power from the Fleischmann and Pons bottle. The two scientists came in for a lot of heavy criticism and had a very hard time.

In March 1990 a report was published in *Nature* of the results of a five-week period of monitoring of the experiment that had been done in Pons's own laboratory by other scientists at the request of the Utah

University authorities. There was, they said, no evidence of fusion activity. An editorial in the same issue of *Nature* suggested that there were not enough scientists with the courage to stand up and explain why they thought cold fusion was nonsense. An attorney for Pons then wrote to the principal author of the *Nature* report demanding that he retract the report or face 'legal action'. This foolish attempt to restrict freedom of scientific free speech by a lawyer, who was later found to be on the payroll of Utah University, caused a storm of protest among the scientists, and two months later the attorney apologized in writing.

In January 1991 Pons resigned his chemistry chair to give more time to cold fusion work and soon both scientists broke off relations with Utah's National Cold Fusion Institute. The institute itself closed down in June when the last of the money voted by the State had been spent. But cold fusion refused to go away. In July 1991 an international cold fusion conference was held in Como and was attended by about 200 scientists from around the world, including Fleischmann and Pons. Most of the participants seem to have been partisans of cold fusion and a report suggested that many held that the work of the two scientists was genuine; that they were being persecuted by institutions such as the American Physical Society and *Nature* and by 'scurrilous journalists'; that top scientists had confirmed the 1989 results of Fleischmann and Pons; and that these had certainly been nuclear events.

The scientific establishment, on the whole, remained sceptical. Soon the main centre for cold fusion research shifted to Japan and a third international conference was held in Nagaya at the end of 1992. There were about 350 participants from 16 countries. The term 'cold fusion' had become pejorative and some supporters preferred a title such as 'chemically stimulated nuclear reactions'. Many felt that a new science had been born. Fleischmann and Pons were now working in France at a Toyota-affiliated research institute. In May 1993 they published a paper claiming that continuous temperature measurements showed that their equipment was producing about four times as much heat as could be accounted for by the electric current they were using. They were no longer claiming, however, that a nuclear reaction was involved. A fourth conference was held in Hawaii at the end of 1993 and about the same time the University of Utah licensed its patent rights to a firm concerned with research into alternative sources of energy. This firm was entitled to exploit any commercial value they might find in the technology. The licence was sold for a six-figure sum.

Although orthodox scientists now almost universally view the claims of Fleischmann and Pons with, at best, scepticism, and, at worst, derision, it is still impossible to say categorically that they blundered in the interpretation of their scientific work. That still remains to be seen, and from what has been said the reader can make up his or her own mind. The millions of pounds of research money spent on the project, especially in Japan, indicate that there were some people who believe that the two men had not blundered. They may have made a discovery of quite fundamental importance. Whether they have made a discovery of the practical importance suggested by the initial hype seems increasingly unlikely. What has become clear is that they, perhaps unwittingly, blundered in the way they presented their finding to the world. No doubt they were intensely excited and felt they could not wait to go through the normal peer-review process. However, by calling a press conference, they inevitably produced implications of large promises that, at least so far, have not been fulfilled.

No doubt these two scientists thought of another pair of previously unknown science workers – James Watson and Francis Crick. Perhaps they felt that their discovery was in the same class as that of the earlier American and English pair. Perhaps if they did, it would have been better for them to have presented their findings in the way Watson and Crick did in April 1953 – by a short paper, not much more than a page in length, that modestly started 'We wish to suggest . . .' and concluded 'It has not escaped our notice that . . .'

There is a fascinating postscript to this extraordinary story. It is now possible to state, with sober literal truth, that cold fusion has been achieved. A Japanese research team, headed by Kanetada Nagamine and working at the Rutherford Appleton Laboratory in Oxfordshire, are fusing atomic nuclei – deuterium (one proton and one neutron) and tritium (one proton and two neutrons) – at a rate of half a million a second and releasing enormous amounts of energy.

There is, it is hardly necessary to state, a snag. When we say 'cold' fusion, we don't mean room temperature; we mean really cold – temperatures near to absolute zero ($-275.15°$C). The atoms that are being fused, having one proton each, normally have one electron each. At the Rutherford Appleton Laboratory, they have a very special particle accelerator that can cause collisions that produce particles called muons. These have the same negative charge as electrons, but are very much heavier than electrons – more than 200 times as heavy. If a stream of muons is fired at liquid or solid deuterium or tritium – hence the need for extreme cold – some of their atoms will naturally pick up a muon instead of an electron. As far as the atom is concerned, it's the negative charge that counts. But the resulting muonic atoms are very different from ordinary atoms. Because of the greater mass of the muons, they orbit much closer to the nucleus than normal and the atoms are, consequently, much smaller and tighter. This means that muonic atoms can come much closer to each other than normal atoms.

A muonic tritium atom can form a molecule with an ordinary deuterium atom. When this happens, the two nuclei are so close together that random fluctuations in their position allow the nuclei to fuse. When they do so, an alpha particle (two protons and two neutrons) is formed, the spare neutron is ejected, and the muon is freed to replace the electron in another tritium atom and achieve another fusion. Each fusion releases a great deal of energy, but, of course, considerable energy is required to produce muons in the first place. The calculations show that the system can produce more energy than it consumes if a muon chain reaction achieves 300 fusions in the lifetime of the muon.

Unfortunately, muons have a very short life – around a millionth of a second – after which they turn into electrons. So the muon has to be busy. Each fusion is very quick – a thousandth of the lifespan of a muon – and the molecule of muonic tritium and deuterium can also form very quickly. The real difficulty lies in the nature of the created molecule. This, of course, has two positive charges (because of its two protons), and so it has a strong tendency to grab a muon (negatively charged) so that it is prevented from proceeding with the chain reaction.

This, so far, is the stumbling-block that has prevented the Japanese team from changing the world and from letting their financiers get back some of the money they wasted on the Fleischmann and Pons disappointment. The Nagamine team have achieved around 200 sequential fusions and are hoping to reach the magic number of 300. This will prove that cold fusion really works. It will not, however, prove that the method is commercially viable. Present calculations suggest that, to achieve this, they will need sequences of 900 fusions per muon.

All the indications are that a controlled fusion source of atomic power will eventually be achieved, and, when it is, the world will be a better place for all of us. However, dreams of doing this on a shoestring in the manner of Fleischmann and Pons are likely to remain what most scientists reckoned them to be from the beginning – dreams.

Mathematics and Computing

Euclid got it wrong

Unfortunately we have few details of the life of Euclid, who was one of the outstandingly great men of all time. He lived around 300 BC and, for over 20 years, worked and taught in the great library at Alexandria, which is said to have held 700,000 volumes. We know from the testimony of others that Euclid was a man of very pleasant personality, modest, fair-minded, generous and kind. His students worshipped him. His writings also show how ready he was to acknowledge the achievements of others and how unassuming he was about his own.

Euclid was very serious about his work. When one of his students asked: 'Is there any money in this?' Euclid said to his servant Grumio: 'Give him a coin. He thinks learning is only for financial gain.' Again, we are told by Proclus (*c.*410–485) that when King Ptolemy I of Egypt, the enlightened ruler who had founded the Alexandrian library, asked Euclid for short cuts to learning geometry that might spare him from having to work through all the propositions from the beginning, Euclid replied: 'In this country, Sir, there are two kinds of roads – the hard road for the common people and the easy road for royalty. Unfortunately, there is no royal road to geometry.' It was typical of Euclid's systematic way of thought that each successive theorem should depend on what had gone before, so skipping was not to be allowed. Like many scholars of the ancient world, he was a man of parts and engaged in scientific studies other than mathematics. He left works on conic sections, astronomy, optics, mechanics, fallacies and the physics of music. Unfortunately, some of these, including the work on fallacies, have been lost. It may be that this book annoyed someone by demolishing long-cherished delusions.

Euclid was a man of great ingenuity. One day during discussion with his fellow teachers he was told that it was impossible to measure the height of the Great Pyramid. 'Really?' said Euclid, 'I don't agree. Try this one. At the time of day when your shadow is exactly the length of your height, measure the length of the shadow of the pyramid. That is its height.' One of the chapters in the book on optics effectively debunks the ridiculous Epicurean insistence that the size of an object is the size it looks, however distant.

Euclid's writings on mathematics – the *Elements* – include more than just plane geometry. They consist of thirteen books covering just about everything that was known of mathematics at the time. They include arithmetic in the sense of the theory of numbers, geometrical progressions, irrational numbers, a geometrical equivalent of algebra, and solid geometry. The work on plane geometry – popularly known simply as 'Euclid' – is the best-selling mathematical textbook of all time, and formed the basis of the mathematical education of people all over the educated world for 2000 years. The book came down to us through Arabic translations which were translated into Latin in the twelfth century and subsequently, in the sixteenth century, into the vernacular. It was first printed in 1482 in Venice, and at least 1000 editions have appeared since then. There can be very few authors who have achieved a record like that.

This book is, by any standards, masterly. The logical methods he used, starting with self-evident statements (axioms) no one could object to, and then moving on to propositions, with a clear demonstration of each stage, until a compelling proof was reached, have, over the ages, had a profound effect on human thinking. Indeed, it has been suggested that Euclid's Elements has had a greater effect on the human mind than any book other than the Bible. His method has deeply impressed many. The English philosopher Thomas Hobbes, on reading 'the sum of the squares on the two sides containing the right angle is equal to the square on the other side' (Pythagoras' theorem), exclaimed, 'By God, this is impossible!' But when he read the proof, he admits that he 'fell in love with geometry'.

Euclid starts with a list of 23 'definitions'. There are some problems with these because they do not always use terms that are better known than the things they are defining. For instance, an angle is defined by saying that it is 'the inclination and meeting of two lines in a plane but not in a straight line'. The objection here is that the word 'inclination' is no better known than the work 'angle'. Some of the definitions are logically circular. He says, for instance, that 'a straight line is a line that lies evenly along all the points on it'. After the definitions come the axioms – statements that everyone can agree on, such as that things equal to the same thing are equal to one another or that the whole of a thing is greater than a part of it. Based on these axioms then come the theorems.

Euclid made the very reasonable assumption that space was flat, and on this assumption drew one or two conclusions that have since been shown to be importantly wrong. He assumed, for instance, that parallel lines, however far extended, could never meet. Euclid just took this for granted and never tried to prove it. He was never quite happy about this, however, and tried to avoid using the parallel lines postulate if he could. Later mathematicians also tried to prove the truth of the statement and failed. Eventually, the great German mathematician Carl Friedrich Gauss (1777–1855) concluded that such attempts were futile. He privately decided that geometries other than Euclid's were possible, but this heretical notion he kept to himself.

One of the later people to try to prove the parallel lines postulate was the young Russian mathematician Nicolai Ivanovich Lobachevsky (1793–1856). This attempt was to have momentous consequences. The work that arose from it was brilliant but was ignored and ridiculed. It was not until after he was dead that the mathematical world woke up to the fact that Lobachevsky had, single-handed, created a whole new geometry that proved that Euclid did not represent absolute truth and that forced them to change their basic ideas of the nature of mathematics.

Lobachevsky decided to see what would happen if he denied the truth of the postulate that parallel lines could never meet. What happened was a complete and valid alternative geometry with no logical contradictions. Lobachevsky published three accounts of this new non-Euclidean geometry and one of these was read by Gauss.

Lobachevsky's idea, and that of several other mathematicians who worked entirely independently of him, notably János Bolyai (1802–1860), was to see what would happen if, instead of considering that geometry applied only to a flat plane, one were to extend it to curved surfaces. If you draw a triangle on the upper concave surface of a horse's saddle, for instance, you immediately deny one of Euclid's theorems – that the angles of a triangle always add up to 180°. On this kind of surface, the angles always add up to *less* than 180°.

Lobachevsky's work was taken further and generalized by the German mathematician Georg Friedrich Bernhard Riemann (1826–66). The lecture in which Riemann first announced his ideas came about in an unusual manner. After graduating from the University of Göttingen, Reimann, a poor, unhealthy and painfully shy young man applied for a job as a *Privatdozent*, a lecturer who lived on fees paid by his students. Anyone who applied for such a job had to show his suitability by giving a lecture before the university faculty. The elderly Gauss was a member. It was normal for three subjects to be offered from which the faculty would choose one or other of the first two. The third was rarely if ever selected but three had to be prepared. Riemann handed over his list, the first two subjects of which he had carefully prepared and knew inside-out. The third he had not prepared, but since he had to give a third subject he had casually put down 'The hypotheses underlying geometry'.

To Riemann's horror, Gauss insisted on the third subject. And, of course, Gauss was much too important to be denied his whim. So Riemann was stuck with this fortunately very general subject. As he warmed to his task, Riemann's halting exposition gradually warmed up and he soon found the words tumbling over each other. The truth was that, having given the matter a great deal of private thought, he had a great deal to say on it. He dealt with spherical geometry and explained that geometry could involve any kind of surface and that the nature of the surface determined the geometry. He pointed out, for instance, that on the surface of a sphere the angles of a triangle always add up to *more* than 180°. Moreover, if you consider the surface of the earth, you will see that, although the meridians of longitude all cross the equator at right angles, and must thus be parallel, they do not stay the same distance apart but converge and meet at the poles. Riemann even went into the topic of curved space and demonstrated that it was possible to construct a four-dimensional geometry that was just as intrinsically consistent and as logical as Euclid's three-dimensional geometry.

This was not just sophistry. Riemann's non-Euclidean geometry led to the concept of curved space and was used by Einstein as a basis for his general theory of relativity. Curved space is very hard to visualize and some notable thinkers have confessed that they can't really do it. Analogies can help and one of the most useful is the idea of two-dimensional beings living on the surface of a sphere or other body (see Newton got it wrong). These beings are free to move as much as they like. Since they can move as far in any direction as they wish, their space is unbounded, but we can readily see that it is finite.

Geometry confined to the surface of a sphere is non-Euclidean. In it, the shortest or straightest distance between two points is not a straight line; it is the arc of the great circle that passes through them. This is how airline administrators plan their flight routes on the earth. In such a system every great circle goes right round their universe and can be divided by two points into two parts. Unless the two parts are equal (because they are at the ends of a diameter of the sphere), the shorter part is undeniably the smallest distance between the two points. But the larger part of the same great circle is the straightest line. So, contrary to Euclid, the shortest line is not the same as the straightest line. In this geometry what seem to be parallel lines always meet in two points. The sum of the three angles of a triangle is greater than 180° and the excess over 180° increases as the size of the triangle increases. This makes it impossible to compare 'similar' triangles of different sizes. Triangles of different sizes can't have angles of the same size. So what became of the relationship between the idea of size and the idea of shape? Relativity rules.

You can readily visualize other geometries based on surfaces shaped in different ways. Suppose we consider a geometry based on a surface like that of an egg. In such a case the rules change seriously if the figures are moved to different parts of the surface. Try moving a triangle about on the surface of such a shape and you will see that its angles are constantly altering. So what becomes of the relationship between the ideas of shape and position? Again, relativity rules. Although we live on the surface of a sphere, our day-to-day geometry is not spherical. The surface of the earth is so very large compared with our Euclidean geometrical figures that Euclid works fine for most practical purposes. However, this is not the case for people planning international flights or for cosmologists, who work on an immensely larger scale.

You may be wondering what all this has to do with science. Well, it has a great deal. If triangles can't be relied on to remain constant in curved two-dimensional space, then solid objects and distances can't be relied on to remain constant in three-dimensional space. Einstein showed – and it was later experimentally proved – that light is deflected by the gravitational effect of a massive body. Einstein in his general theory of relativity insisted, however, that light beams are straight and that it is space that is curved. Riemannian geometry can precisely calculate the curvature of space caused by a massive body. The very nature of gravitation depends on the existence of geometries that are non-Euclidean.

So it seems that Euclid's geometry is only just one special case of the class of all geometries. It is the case in which the curvature of space is zero and in which everyone accepts Euclid's axioms. In the wider context of science it is also, paradoxically, the least important of all the possible geometries. Euclid, in his modesty, would have been the first to agree.

> Aerial flight is one of that class of problems with
> which men will never have to cope.
>
> *Simon Newcomb (1835–1909),*
> *the most influential American astronomer of his day*

Any fool can use the differential calculus

The physicist Silvanus P. Thompson (1851–1916) was a Fellow of the Royal Society and professor of applied physics and electrical engineering at the City and Guilds Technical College at Finsbury. He was president of the Institution of Electrical Engineers, President of the Physical Society, and President of the Optical Society. He had a BA and a BSc and was a Doctor of Science, all from London University. He made important contributions to the study of optics and electricity, wrote many books, some of which became long-term standard texts for electrical engineers, and his public lectures attracted large audiences. He wrote excellent biographies of Michael Faraday, Lord Kelvin and Philipp Reis, and translated the seminal work *De magnete* of the sixteenth-century scientist, William Gilbert. He was a talented painter, and produced some excellent paintings of Alpine scenery.

Altogether, you might think, quite a bright specimen. Well, we have it from Thompson, himself, that he did not agree. This is what he wrote as a prologue to his excellent little book *Calculus Made Easy:*

'Considering how many fools can calculate, it is surprising that it should be thought either a difficult or a tedious task for any other fool to learn how to master the same tricks . . . The fools who write the text-books of advanced mathematics . . . seldom take the trouble to show you how easy the easy calculations are. On the contrary, they seem to desire to impress you with their tremendous cleverness by going about it in the most difficult way. Being myself a remarkably stupid fellow, I have had to unteach myself the difficulties, and now beg to present to my fellow fools the parts that are not hard. Master these thoroughly, and the rest will follow. What one fool can do, another can.'

The Republic has no need for men of science.
*Jean-Paul Marat (1743–93), on the occasion of condemning
the celebrated chemist Lavoisier to death at the guillotine*

The destruction of Alan Turing

One of the outstanding names in the history of computing is that of Alan Mathison Turing (1912–54). Turing inherited good scientific genes and came from a family distinguished for its engineers, three of whom had become Fellows of the Royal Society. His father, however, was a member of the Indian Civil Service and Alan was sent to school at Sherborne while his parents were abroad.

While a graduate mathematics student at King's College, Cambridge, Turing wrote a paper that was so brilliant that he was immediately elected a Fellow of his college. In 1936 he went to Princeton University to work on mathematical logic and wrote a paper that proved that there were mathematical problems that could not be solved by a purely mechanical process – in other words, by a machine. In the course of this paper Turing described an imaginary machine that had a large data store, a program that specified what was to be done with the data, and a step-by-step procedure. At the time, few people paid any attention to this. In fact, what Turing had done was to lay down the basic structure of the digital computer of today.

Fortunately, Turing's subsequent career allowed him to implement this kind of reasoning in practice. He returned to Cambridge and, with the outbreak of World War II, he was employed by the government at the Code and Cipher School at Bletchley Park to work on ways of breaking the German cipher codes. In this he was remarkably successful, being centrally concerned in the design of the Colossus machine that was used to break the German Enigma code. Colossus was a digital computer although not a general purpose machine. The results of this work of Turing's enabled the British Navy to intercept and attack German U-boats and to interfere with their maintenance supplies. It contributed notably to the outcome of the war, saved countless British lives, and earned Turing the OBE.

At the end of the war, Turing went to the National Physical Laboratory (NPL) at Teddington, where a machine computation department had just been set up. He wasted no time. In early 1946 he presented a design for a computer, with an estimate of cost (£11,200) to the executive committee. Turing was not just a theoretician. He knew that theoretical ideas had to be reflected in hardware if progress was to be made. So he spent hours linking up valves, resistors, capacitors and inductors with a network of wires on a 'breadboard'. These 'bird's nests' of components held together with blobs of solder were typical of the time. Turing was trying to implement logical gates – devices that gave a one pulse (true) output or a 0 pulse (false) output as required, from various combinations of inputs.

These gates are at the heart of digital computing. Nowadays, of course, they are all sealed away in integrated circuit chips based on low-voltage transistors, but Turing had to work with thermionic valves that required up to 500 volts on their anodes. A colleague describes how he would watch Turing poking his hot soldering iron into such lethal contraptions. Turing's design consisted of both the theoretical basis and the practical physical design for what was to become one of the first real digital computers, the automatic computing engine (ACE).

The director of the laboratory, Charles Darwin (the grandson of *the* Charles Darwin), was very doubtful whether mathematicians like Turing would know enough practical engineering actually to make the ACE and he decreed that the radio division of the NPL should get into the act. Inevitably, the people in the mathematics division were annoyed. Turing went back to King's for a year's sabbatical and the work went on without him. A trial version of ACE was eventually completed and was a brilliant scientific success. It was very fast and was able to perform much useful work. It came to be used regularly in the NPL mathematics division, where the staff were encouraged by its success to write programs for it. This was really how programming started. Four hundred ACE machines were sold in the USA.

Unfortunately, as was so often the case in the history of computing, the machine was made using technology that quickly became obsolete. It relied heavily on acoustic delay lines – devices using sound pulses circulating in tubes full of mercury – and had the arithmetic and logic systems built into the delay line storage. The programs were stored in the delay lines. Computers need to be able to hold data temporarily while other parts of the program are performed. Even more significant for the downfall of the ACE was the development of the transistor during the time that the machine was being built and developed.

In 1948 Turing became one of the first to hold a university appointment in computing. He was appointed Reader in the theory of computation at Manchester University and was assistant director of the project that built the Manchester Automatic Digital Machine (MADAM).

By this time, the appearance of these 'electronic brains', as they were then popularly described, had aroused the highly emotive question: can machines think? Turing had strong opinions on the matter. In 1950 he published a paper in the psychological journal *Mind*, in which he suggested the following test: put a computer in one room and a person in another and arrange two-way teletype communication between them. Run a program so that the two can conduct a conversation, and arrange it that both contributions can be studied by an observer. Turing claimed that, given enough computing power and a suitable program, it should be impossible, at least for a reasonable time, to determine which room contained the human and which the machine. This is the famous Turing test of artificial intelligence. The point about it is that if a setup is achieved that passes the Turing test, we cannot logically attribute intelligence to the human without ascribing it to the machine.

Turing was elected a Fellow of the Royal Society in 1951. The following year, when he was at the height of his powers and only 40 years old, he was arrested and tried for a homosexual offence with a young Manchester print worker with whom he was having an affair. He was convicted and was offered the alternatives of going to jail or accepting hormonal medical treatment to cure this 'disease'. That was the state of the law, and even of scientific opinion on the matter, at that time. Turing accepted the medical option, although he must have known that treatment purporting to destroy or reverse his normal sexual impulses was a travesty of reason and justice.

He lived through this shameful martyrdom for two years, seldom leaving his house. Then he took cyanide.

I had the idea of a new kind of pen that used a ball instead of a nib. But I decided it wouldn't work, so I dropped the project.

Chester Carlson (1906–68), American inventor of the Xerox copier that made him a millionaire

The millennium date blunder

It seems hardly credible that so few people could have foreseen the problem. Considering that the enormous preponderance of software has been written within a decade of the turn of the century, it is remarkable that so many programs had a ticking time bomb built into them. Or it *seems* remarkable now that we are all aware of the problem. Another brilliant piece of hindsight.

Come 1 January 2000, millions of computers all over the world are going to be in trouble. This is because, for decades, it has been entirely satisfactory for us to record and store the figure for the year as two numbers. For example, 10 March 1995 has been stored as 10/03/95 and everyone, and every computer, has understood exactly what that meant – except, of course, in the USA, where it means 3 October.

Using two numbers for the year worked fine before computers were invented and everyone knew which century they were in. It also works fine throughout the computer era of the twentieth century, right up to 1999. However, if you have told your computer that 99 means 1999, you must not be surprised if, when your computer moves on to the following year and looks at the figures 00, it thinks it is in the year 1900. If millions of different items of data have been stored using this unimaginative convention, then there are going to be problems.

Suppose, for instance, we are talking about a computer in a bank, and suppose that this computer is set up to calculate the interest on your deposit account. No problem. All the machine has to do is to subtract the date at the start of the period from the date at the end of the period and multiply this by the interest rate the bank has promised you. Suppose you have £100,000 in your deposit account and you are getting 6 per cent interest per annum. If the period is a year, the bank will credit your account with £6000. Or will it? Not if it has to subtract 99 from 00. The result of this calculation could well turn out that, since you are due £6000 for minus 99 years, you owe the bank £594,000. To say nothing about compound interest.

Remember that computers don't have emotions. So the bank's machine will coldly inform you of your sins and may include a programmed dire warning as to the consequences of running up an unsecured overdraft. Of course, you could get lucky. The computer might simply ignore the minus sign and *credit* you with 99 years' interest. These fanciful consequences of the millennium blunder are unlikely, but other, even more serious, consequences are inevitable unless businesses get their act together in plenty of time. One possible disastrous consequence is that irretrievable data may be overwritten with nonsensical figures. The backup tapes of discs, with the previously saved correct data, may also be overwritten. Chaos. And a wonderful opportunity for the crooks. Furthermore, 1 January 2000 is a Saturday, and, being the millennium, there are bound to be additional public holidays, so there will be plenty of time for all the records to be overwritten with garbage.

The essence of the problem is that almost all business software, whether used by large institutions, small firms or private individuals, is date-dependent. The proper and honest management of insurance premiums, mortgages, invoices and orders all depend on the computer being able to manipulate dates correctly. But it is not only business software that is involved. Public utilities of all kinds, the maintenance and provision of supplies, control of shipping and power stations, defence systems, weather forecasting, and many more essential services may be affected by the millennium date blunder.

Even more serious is the fact that lives depend on computers. Suppose, for instance that a crew of shuttle astronauts were on a mission that started late in December 1999 and continued over the new year period. Further suppose that vital computer systems were date-dependent, as many are. The result could be a disastrous shutdown of life-support systems. Some medical life-support systems also rely on computers and may be date-dependent. Let's hope that by the time this book is published the most serious of the problems arising from this all-too-human lack of imaginative insight will have been sorted out. But it seems improbable.

Chemistry

Fermentation is a purely chemical process. If it were possible to recombine the alcohol and the carbonic acid gas, the sugar would be recovered.

Antonie Lavoisier (1743–94), the French chemist and one of the founders of modern chemistry

The four elements

In the early stages of scientific thought, fundamental ideas were necessarily speculative and it may seem a little unfair to categorize one of the most important of these as a scientific blunder. Nevertheless, the consequences of this particular mistake were so massive, so serious and so persistent – they are discernible to this very day – that this must be rated as one of the greatest scientific blunders of all time. Chemistry is the study of the what the world is made of and this is the story of an early theory of chemistry and its disastrous consequences.

On the south coast of Sicily, on a hill overlooking the Mediterranean Sea is the town of Agrigento. Two and a half thousand years ago this town was called Agrigentum and it was one of the most beautiful and prosperous cities in the ancient world and a centre of Greek culture and sophistication. In one of the most prominent and wealthy families of the town, lived a remarkable man called Empedocles. He was famous. He was a doctor, physicist, poet, philosopher, social reformer and statesman. He was also a con man who claimed to foresee the future. In those days they were called soothsayers (truth-tellers).

Agrigentum was a lovely place to live in if you were rich. The sea view was magnificent and the surrounding country fertile and pleasant; it also contained hot springs, and mines of salt, gypsum and sulphur. Farmers brought in their produce to the town market, as they do today. At the time, Sicily was a Greek colony. Empedocles could afford to travel and his natural interests took him eastward to Greece where intellectual activity was high and ideas were in a ferment. So he had plenty to think and write about, which he did in the form of long poems. Unfortunately, most of his written work has disappeared, but we still have numerous fragments amounting to some 450 verses.

Empedocles was a real scientist and interested in everything. He proved by experiment that air was a real substance and not just empty space. He worked up some ideas about the nature of the universe. He speculated on the nature of vision and on how objects are visible to us. He even decided, correctly, that light moved with a definite speed. He was very interested in medicine and showed the importance of the arteries and veins. But his most important idea – and the one that gets him into this book – was that everything in the material world was made up of four elements. These elements, or roots, were earth, air, fire and water. That was all there was to it. The four elements were driven by two forces – love that brought them together and strife that drove them apart. Empedocles also applied the four-elements theory to medicine. Health was the result of a proper balance of the four elements in the body; disease resulted if they got out of equilibrium.

The idea of the four elements caught on and was soon universally believed. It was elaborated to include four qualities – moist, dry, hot and cold. Water was moist and cold; earth was dry and cold; fire was hot and dry; and air was moist and hot. This idea made it even more universally applicable. It was perceived as explaining everything in such a satisfactory way that no other basic theory of matter was needed. The idea, of course, could not be disproved until the development of modern chemistry in the eighteenth century and the atomic theory in the nineteenth century. Ideas that can't be disproved are often dangerous. This one certainly was. Such an idea is apt to acquire the status of a religion and to stifle alternative ideas.

In the biological context, the idea of equilibrium in the four elements soon suggested an offspring that seemed to go even further in explaining ill-health – the four fluids or humours (even today the term 'humour' persists in the phrase 'aqueous humour' of the eye). These were phlegm (mucus), blood, black bile and yellow bile. Each of these was perceived as exerting an influence not only on health but also on character. This idea was so strong that words deriving from it are still in common use today by people with a literary turn of mind: phlegmatic (stolid and unemotional); sanguine (cheerful, confident and optimistic); melancholic (depressed); and choleric (angry). Even the word 'humour' in the sense of a disposition to amusement derives from this idea.

The four humours were closely linked to the four elements. Water was associated with phlegm; air with yellow bile; earth with black bile; and blood with fire. So any particular disease could be accounted for by postulating an excess of the particular humour or humours that gave rise to the observed symptoms. A person with a fever and a sore throat for instance would be diagnosed as suffering from a general excess of heat and dryness of the throat. Heat and dryness were the qualities associated with blood, so a person with these symptoms had an excess of blood and had to be relieved of it. This is why blood-letting, by opening a vein and letting blood run away or by the application of blood-sucking leeches, was such a prominent feature of medical practice in the 2000 or so years from the time of Empedocles.

This scheme of medicine was most powerfully promoted by a Roman physician called Claudius Galen, who lived some 600 years after Empedocles but who took up his scheme with enthusiasm. Galen's influence was enormous and persisted for 1500 years. No other medical system got a look-in. Unfortunately, there was not much in a practical way that could be done about an excess of any of the fluids (humours) of the body other than blood. Purgatives and emetics were commonly used and there were drugs called cholagogues that were supposed to increase the flow of bile into the intestine. But the emphasis was always on blood-letting. This was performed for any disorder featuring a fever or for any condition in which the patient was red-faced or was, for any other reason, thought to have too much blood. The result was that millions of people, many of whom desperately needed all the limited blood they had, were killed by the doctors.

Nutritional anaemia was very common in early societies and the last thing most people needed was to be deprived of blood. If the first bleeding did not cure the patient, this was usually interpreted as indicating that not enough blood had been taken, so the patient was bled again. If, in spite of all the efforts of the doctors, the patient died, this was simply because the disease was too severe. Galen had prescribed blood-letting and Galen could not be wrong.

The real blunder in all this is to base an important system on pure imagination. The idea that everything is made of earth, air, fire and water is nonsense. The four humours as the determinant of disease and personality is nonsense and, of course, the conclusions that were derived from this ingenious idea were complete and dangerous nonsense. The totality of human distress and misery that resulted from Empedocles' seemingly harmless and ingenious idea beggars the imagination. <u>Few disasters in history can compare with it.</u>

Even in Empedocles' time there were people who recognized that science should be based on observation and experiment, not on the assertions of a pundit, however impressive. No one listened to them. Therein lies the blunder.

The history of science shows that important original work is liable to be overlooked, and it is perhaps the more liable the higher the degree of originality. The names of Thomas Young, Julius Mayer, Nicholas Carnot, J. J. Waterston and Balfour Stewart will suggest themselves to the physicist; and in other branches, doubtless, similar lists might be made of workers whose labours remained neglected for a shorter or longer time.

The great English physicist Lord Rayleigh (1842–1919),
President of the Royal Society,
speaking at the 1897 anniversary meeting

The historical oxygen blunder

'Every schoolboy knows,' as Macaulay used pompously to assert: 'that Montezuma died in 1520 during the Spanish conquest of Hernando Cortés.' A latter-day Macaulay might, perhaps with more justification, claim: 'Every schoolboy knows that oxygen was first discovered by the English chemist Joseph Priestley.' Undeniably, this statement is to be found in all the elementary chemistry books and it is taught in every elementary chemistry course. The chances are that most chemistry teachers think it is true.

Oxygen, of course, is the one outstandingly vital element. We can survive for weeks without food and for days without water, but if we are deprived of oxygen for just a few minutes we are dead. Oxidation – chemical combination with oxygen – is probably the most important of all chemical reactions. It is the ultimate source of energy for nearly all living organisms. It can be rapid, as in burning, when energy is quickly released, or slow as in the reactions in the cells of the human body using enzymes. To survive, the human body requires a constant supply of oxygen. This is taken from the atmosphere and rapidly distributed by the blood to every cell. The air we breathe consists of about 20 per cent oxygen and 78 per cent nitrogen, with traces of rare gases and about 0.03 per cent carbon dioxide. Nitrogen is an inert gas, and is largely ignored by the body; it becomes important only when the body is exposed to such high pressures that the gas dissolves in the body fluids in large quantity.

Joseph Priestley (1733–1804) was an outstanding scientist but remarkably wrong-headed. Oddly enough, although the chemical theory to which he subscribed throughout his life – the phlogiston theory of combustion – was eventually refuted by his own efforts, he never abandoned it. His contributions to science were, however, considerable. He was especially interested in gases. The gas given off in brewing – now known as carbon dioxide – fascinated him, so he collected a good supply of it and showed that, under pressure, it would readily dissolve in water. This was the start of the soda water and soft fizzy drinks industry, and of the wrongly-titled 'aerated water' craze. Almost any bland fluid, including flat water, can be made highly palatable by dissolving carbon dioxide in it. Fortunes are currently being made by converting 'natural spring water' into a tasty drink by dissolving carbon dioxide in it.

The phlogiston theory, in essence, was that anything that could burn or support combustion contained a special substance called phlogiston that was ejected during burning and lost into the atmosphere. Priestley did may experiments, heating metallic oxides (ores) with a large magnifying glass to focus the sun's rays and observed the properties of 'phlogiston'. In 1772 he discovered that plants gave off a gas that was necessary to animal life. Two years later he produced samples of the same gas by heating red oxide of mercury and red oxide of lead. He tried breathing this gas himself and found that it was harmless, producing only a rather pleasant feeling of lightness in the chest. When he compared the new gas with ordinary air in separate sealed containers of equal volume, he found that the new gas would keep a mouse conscious twice as long as air. Priestley concluded that he had discovered a new kind of air from which phlogiston had been removed, and which he entitled dephlogisticated air. The fact, of course, was that he had discovered oxygen.

The Swedish apothecary and chemist Karl Wilhelm Scheele (1742–86) is almost unknown, but is now gradually becoming recognized by the historians of science as probably the greatest chemist of the eighteenth century. He was a remarkable man, with a range of chemical discoveries to his credit that is almost unrivalled in science. Scheele devoted his life exclusively to the pursuit of science, and probably shortened it by his habit of tasting all the new substances he discovered. He isolated the elements arsenic, barium, chlorine, manganese, molybdenum and nitrogen, and produced scores of important new compounds, including barium oxide, benzoic acid, citric acid, copper arsenite (Scheele's green), gallic acid, hydrogen cyanide, hydrogen sulphide, hydrogen fluoride, lactic acid, malic acid, oxalic acid, various permanganates, silicon tetrafluoride, tartaric acid, tungstic acid and uric acid. In fact, his contributions to chemistry probably exceeded those of any other scientist of the time.

In 1775 he was elected to the Royal Academy of Sciences of Sweden – a unique honour for an apothecary's assistant.

Scheele also discovered that certain silver salts were modified by light, thereby anticipating the basis of photography by 50 years. He drew attention to the remarkable similarity in structure between graphite and molybdenum disulphide (now known commercially as Molyslip), since when the latter has been widely used as a better solid lubricant than graphite.

In 1771, while heating mercuric oxide, Scheele discovered a new gas with remarkable properties. It was colourless and odourless and small animals such as mice kept in an atmosphere of this gas became frisky. A glowing wood splint plunged into it would burst into flame. Scheele called the gas 'fire air', and was the first to show that air consisted of two gases, one of which would support combustion, and one of which would not. Scheele immediately wrote a book called *A Chemical Treatise on Air and Fire* in which he described his experiments. His neglect by the historians of science is largely due to the idleness of his patron the famous Swedish chemist Torbern Bergman (1735–84), who had undertaken to write the introduction. Scheele, a quiet modest man, was probably too timid and too interested in other research to think of nagging Bergman. Be that as it may, Bergman took so long to write the introduction that Scheele's publisher (who has occasionally been blamed for the delay) was unable to get the book out until 1777. By then, of course, Priestley had long since reported his own similar experiments, and had taken the full credit as the discoverer of oxygen.

This was not quite the end of the story. In 1774 Priestley visited the great French scientist Antoine Laurent Lavoisier (1743–94), and told him about his experiments. Lavoisier, a rather arrogant man who regarded Priestley as an amateur, immediately tried out Priestley's experiments and confirmed the finding. He at once realized the importance of the discovery and saw that the new gas was the essential ingredient of air. After a bit of muddled thinking he decided to call it oxygen – the Greek for 'acid maker' (which it is not). Lavoisier was convinced, wrongly, that all acids contained the newly discovered gas. He did, however, like Scheele, show that air contained two main gases, one which supported combustion (oxygen) and one which did not (nitrogen). He also studied the heat produced by animals breathing oxygen, and demonstrated the relationship of respiration to combustion.

All this was so important to science that Lavoisier became determined to go down in history as the discoverer of a new element. So he claimed the discovery as his own and did not acknowledge the help he had had from Priestley. Scheele, who was the real discoverer of oxygen, was, of course, ignored and forgotten.

Pasteur's anti-rabies treatment is useless, dangerous and devoid of any scientific character. Pasteur is not curing rabies, he is actually communicating it. His laboratory should be closed.

Professor Peter of the French Academy of Medicine and one of Pasteur's most determined opponents, speaking after the news that Pasteur's vaccine had saved the life of the boy Joseph Meister who had been bitten by a rabid dog

---

A pathological modesty

Henry Cavendish (1731–1810), one of the all-time most distinguished researchers in physics and chemistry, came as near to being a mad scientist as any in recorded history. He was also a millionaire who spent hardly a penny of his vast inherited fortune. Cavendish was the son of Lord Charles Cavendish and the grandson of the Duke of Devonshire on his father's side. His mother was Lady Anne Grey, daughter of the Duke of Kent. Such refined aristocracy may have had something to do with the grossly reclusive habit of life he adopted from an early age.

He was educated at the Reverend Dr Newcome's Academy in Hackney and, for over three years, at Peterhouse, Cambridge, which he left without taking a degree. Such a distinction was considered scarcely necessary to someone in the situation in which Providence had been pleased to place him, and this was a great relief to the young man who had been dreading the thought of having to face his professors in *viva voce* examinations. It was not that he feared the exposure of his ignorance – he was a man of brilliant and scholarly mind – it was simply that he could not bear to talk to anyone. In a desperate emergency he might utter a few words to one man, but never to more than one. Most of the time he was silent. Cavendish probably spoke fewer words in the course of a life of nearly 80 years than anyone other than a Trappist monk.

As for the thought of talking to a woman, that was completely out of the question. Cavendish was terrified of women. Of necessity he had to give orders to his servants, but this was done by written notes. His dinner he ordered each day by leaving a note on the hall table. Any servant girl who allowed herself to be seen in his house was sacked on the spot. His voice, although seldom heard, was thin and high and he was always annoyed if anyone looked at him. Occasionally, however, he would quietly approach to hear what others were saying if he thought the subject worthy of note.

Outside science and scientific matters he was coldly indifferent. He was never known to refer to politics or religion and was apt to reject any overtures of human sympathy. As may be expected, Cavendish had a large library and he was generous in allowing his scientific acquaintances to use it. But the library was four miles from his home on Clapham Common and he would never go there if he knew that anyone else was present. When he did go, he would never take a book out without first writing out a formal receipt for it.

Mathematics attracted him early. In 1765 he recorded the results of an extensive series of experiments of heat – discoveries that he never mentioned but which would have greatly advanced the progress of science. Around 1770 he then proceeded to research into electricity and advanced his own knowledge of the subject with a series of discoveries that anticipated much that was to be found out by great men such as Charles Coulomb and Michael Faraday 50 years later. None of this was known until Cavendish's notebooks were read by James Clerk Maxwell a hundred years later. Maxwell, who was by then the greatest theoretical physicist in the world and the author of *The Electromagnetic Theory of Light* (1868), was astonished to find out how much Cavendish had discovered and immediately arranged for the publication of the notebooks. But this was merely of historic interest and, because of Cavendish's extraordinary personality, the world had been denied immensely important advances for more than half a century.

Cavendish discovered a new, very light and inflammable gas and worked out its properties in detail. Fortunately, he was able to bring himself to communicate this discovery – in writing – to the Royal Society. Twenty years later, the French chemist Lavoisier (see The historical oxygen blunder) gave the new gas its name – hydrogen. Cavendish also discovered that when the new gas was burned in air water was produced. This was a finding of primary importance. It showed that water was not an element, as had been believed from the time of the Greek philosophers, but that it was a compound of at least two other substances. The implications of this for chemistry were enormous, but Lavoisier got all the credit.

Cavendish's most impressive achievement was to weigh the earth. Isaac Newton had determined that the gravitational attraction between any two bodies was proportional to the product of their weights and inversely proportional to the square of the distances between them. Cavendish worked out that if he could measure the gravitational pull between two bodies of known mass (weight) and size he could apply this figure (the gravitational constant G) to Newton's equations for the pull between the earth and a body of known weight.

So Cavendish set up a delicate experiment. He took a light rod, 40 inches long, and fixed a two-inch lead sphere to each end of it. Then he suspended the rod horizontally at its centre by a long wire so that the rod could turn freely causing the wire to twist a little. He waited until the rod had stopped turning and then brought two very large balls of metal close to the little balls, one on each side, placing them so that the gravitational pull of each on the small balls would cause the rod to rotate a little against the minute twisting resistance of the wire. Cavendish knew the mass of all the bodies and the exact distance of the small balls from the large balls. Now he also knew the tiny force that had been exerted to twist the wire. All that remained was to do the calculations. The earth turned out to have a mass of 6,600,000,000,000,000,000,000 tons. He also knew the dimensions of the earth, so it was easy to calculate its density. This turned out to be 5.5 times that of water so there had to be a lot of heavy stuff in there somewhere. This turned out to be iron.

Cavendish was so retiring that he was unaware of contemporary developments in science. The eminent French zoologist Georges Cuvier, in his threnody on Cavendish, said of him: 'His demeanour and the modest tone of his writings procured him the uncommon distinction of never having his repose disturbed either by jealousy or by criticism.' He died on 10 March 1810. Having ordered his manservant not to come near him till night time, he remained alone all day in a dying state. The servant found him in this condition and immediately sent for the physician Sir Everard Home. Cavendish said to the doctor: 'Any prolongation of life would only prolong its miseries.' Then he died. He left his untouched fortune of £1,175,000 to his cousin, Lord George Cavendish.

Cavendish was hardly known in his lifetime but is now recognized as one of the great scientists of all time. The Physical Laboratory at Cambridge, from which came advances such as atomic energy, X-ray crystallography and the structure of DNA, is named after him.

> I remember at an early period of my own life showing to a man of high reputation as a teacher some matters which I happened to have observed. And I was very much struck and grieved to find that, while all the facts lay equally clear before him, only those which squared with his previous theories seemed to affect his organs of vision.
>
> *Joseph, Lord Lister (1827–1912),*
> *who introduced antiseptic surgery*

Alchemy's failure

There is a remarkable novel by Balzac called *The Quest of the Absolute*. The central character, Balthazar Claes, is driven for over 20 years by an obsessive quest that ruined him financially, destroyed his marriage, killed his devoted wife and ended in his madness and death. The quest was for the Philosopher's Stone – the wonderful substance that could transmute base metals such as lead into gold. Balzac set his novel in the first half of the nineteenth century, so Balthazar must have been one of the last of the alchemists. By then, science had overtaken this futile and wasteful activity and Claes was a late throwback – a man whose earlier studies had gone sadly astray. He had been reading the wrong books.

Alchemy actually dates from very much earlier and derives from primitive superstitious beliefs that extend back into the mists of prehistory. It had its heyday in Europe, however, between the twelfth and the fifteenth centuries, after Arabic alchemical texts were translated by scholars. The two central concerns were the Philosopher's Stone and the Elixir of Life. Possession of the former meant untold wealth and of the latter immortality. Alchemy was tied up with another pseudoscience, astrology, and with the general notion of perfectibility. The astrologers taught that, given the appropriate astrological influences, imperfect, or base, metals could be 'perfected' and turned to gold.

The general idea behind this was that metals grew in the earth and very gradually, over time, developed from 'base' metals to 'noble' metals, and eventually to gold. Gold was the perfect metal because it never corroded. Normally, perfectibility was a very slow process, but those who were in the secret could hasten the process. Heating a metal for long periods could, for instance, kill it so that it could then be revived in a finer form. This idea of the growth of metals to a nobler form was very widespread. There is a reference to it in Ben Jonson's play *The Alchemist*: '. . . of Lead and other Metals, which would be Gold, if they had the time'.

Jonson's extraordinary play is worth reading for the exuberance of the language alone. Here is a sample. Sir Epicure Mammon, anticipating what the Philosopher's Stone will bring him, says:

> My meat shall all come in, in Indian shells,
> Dishes of agate, set in gold, and studded
> With emeralds, sapphires, hyacinths, and rubies.
> The tongues of carps, dormice, and camel's heels,
> Boiled in the spirit of Sol, and dissolved pearl.
> And I will eat these broths with spoons of amber,
> Headed with diamond and carbuncle.
> My foot-boy shall eat pheasants, calvered salmons,
> Knots, godwits, lampreys: I myself will have
> The beards of barbels serv'd, instead of salads;
> Oiled mushrooms; and the swelling unctuous paps
> Of a fat pregnant sow, newly cut off,
> Dressed with an exquisite and poignant sauce;
> For which I'll say unto my cook, 'There's gold,
> Go forth and be a knight.'

Unlike later science, alchemy was esoteric and hidden and secrets were carefully and jealously hidden. Strange symbolic names or signs were used to conceal the true nature of the materials used. Common symbols related gold and silver to the sun and the moon, respectively; iron to the planet Mars, with the universal male symbol ♂; copper to the planet Venus, with the universal female symbol ♀; tin to Jupiter; lead to Saturn; and mercury to the planet Mercury.

The transmutation of certain elements was achieved in the twentieth century, not for its own sake but as an incidental event in the study of the nature of the atomic nucleus by nuclear physicists. This was advanced science and it is not in the least surprising that the alchemists failed to achieve it. No modern physicist would think of trying to produce gold by transmutation. If this were possible, the cost of so doing would enormously outweigh the value of the gold produced. Their failure drove many alchemists into fraud. Many became accomplished gold-fakers. Those with a spark of honesty claimed that they were able to make a substance indistinguishable from gold, or even better than gold.

Gold-faking, or gold-making from inferior materials, was called aurification. All kinds of methods were used, from the crudest to the subtle. Here are two samples: 'If you want to make a copper ring seem so like gold that it cannot be detected either by the feel of it or by the use of the touchstone, proceed as follows: grind some gold as fine as the finest flour. Do the same with lead. Take two parts of the lead powder to one of the gold. Mix them up thoroughly and then mix with gum. Coat the copper ring with this mixture and heat it. Do this several times until the ring appears to be of pure gold.' The touchstone, by the way, was a hard piece of dark basalt or jasper, or other siliceous stone, used to test the quality of gold and silver by rubbing along it and observing the colour of the mark produced.

A less crude method of aurification used the alloy of gold and silver called 'asem'. 'Take some small pieces of purified soft tin. Mix four parts of this with three parts of pure white copper and one part of asem. Melt them all together and cast whatever object you like.' This was a very sneaky trick because, if the metal was suspected of not being gold, the maker could 'admit' that it was asem, and still cheat. Some of the tricks of the fraudulent alchemists are described by Chaucer in the 'Canon's Yeoman's Tale' in his *Canterbury Tales.* They would, for instance, conceal some gold in the false wax bottom of a crucible so that it appeared on melting, or they might stir the crucible with a fake stirring rod containing a little gold. Charcoal could be prepared with a hollow containing gold and sealed with wax. These and various standard tricks of prestidigitation were used by fraudulent alchemists to persuade the gullible that they could make gold.

The great quest was for the Philosopher's Stone. Here is a valuable account of how to prepare this long-sought magical substance offered free of charge to anyone who wants to try it: 'Take some ordinary gold, some ordinary silver and some ordinary quicksilver (mercury). Separately purify all three, removing all adulterants. Treat the gold with quintessence of sulphur, the silver with philosophical mercury, and the quicksilver with philosophical salt. Now mix all three in the philosopher's egg (the Hermetic Vase). Then wait for the process of the great work, the outcome of which will be The Stone.'

Some historians of science insist that alchemy was the direct forerunner of modern chemistry. They point to the many pieces of laboratory apparatus developed by the alchemists and used later by real chemists, and they list processes of purification and even separation of substances as examples of early chemical analysis. In their efforts to produce gold, the alchemists learned a good deal about the alloying of metals and about the production of various dyes.

There are some serious objections to this theory. The things the alchemists discovered or developed were incidental to their purpose, which was always selfish. In their frantic efforts to make gold, they learned many experimental techniques that later proved useful to science, but these techniques were not developed for scientific purposes. Few if any of the alchemists were motivated by a disinterested quest for knowledge. Their purpose was not to elucidate the nature of matter – to discover what the world was made of. Rather they were driven by greed. Chemistry, as a scientific discipline, began when the motivation for the work changed from the quest for personal advantage to the quest for knowledge.

I do not think that the wireless waves I have discovered will have any practical application.
Heinrich Rudolf Hertz (1857–94), discoverer of radio wave
propagation

Isaac Newton, alchemist

Ask a physicist or mathematician to name three great figures in the
history of their disciplines and the chances are that they will mention
Newton. The great English scientist Isaac Newton (1642–1727)
invented the calculus and discovered the binomial theorem. He
determined the laws of motion thereby producing what was virtually
a new branch of physics. He formulated a theory of planetary
movement, wrote a magnificent treatise on optics, elucidated fluid
mechanics, and produced a theory of gravitation that stands to this
day and that has had a profound effect on the development of science.
Perhaps most important of all, Newton developed a strict scientific
method and showed how physical phenomena could be described in
mathematical terms. His book *Philosophiae naturalis principia math-*
ematica (1687) has been described as the most important scientific
book ever written. His influence on the scientists that followed him is
incalculable.

His intellectual vision was extraordinary. Einstein's comment on
Newton was that, to him, 'Nature was an open book.' Newton's
perception of the reach of science is crystallized in his famous
saying: 'I know not what the world will think of my labours, but to
myself it seems that I have been but as a child playing on the
seashore; now and then finding a smoother pebble or a prettier shell

than ordinary, whilst the immense ocean of truth extended unexplored before me.' Call this false modesty if you like; the important point is that it was an expression of his insight into the way things really are. Newton also said: 'If I have seen further than others, it is only because I have stood on the shoulders of giants who have come before me.'

When Newton was in his early 40s and at the height of his powers he practically abandoned scientific investigation and devoted himself to biblical studies and prophesies, magic and alchemy (see Alchemy's failure). The latter, especially, in which he had been interested since his youth, became an intense preoccupation. He spent a lot of money on apparatus, including a furnace, set up an alchemical laboratory, searched out the writings of earlier and contemporary alchemists, and carried out many attempts to realize the promise of these books and manuscripts. His own writings on the subject amounted to about a million words.

Throughout the 1670s he laboured away in his laboratory, patiently studying these esoteric texts, and endlessly trying the experiments. He was especially anxious to prove the hypothesis that the substance 'child of Saturn' (the poisonous, brittle, silvery-white, crystalline metal antimony) gave off magnetic rays that could attract the life force of the world. It seems that Newton believed that this element carried 'God's signature' (whatever that may have meant).

Now Newton was not mad. His mental powers were undiminished. To him, alchemy seemed to offer the possibility of extension of scientific knowledge beyond what could be achieved by conventional science. In this context, one of the most powerful influences on Newton was the work of the Polish alchemist Michael Sendivogius (1566–1636). In 1604 Sendivogius had published a book *De lapide philosophorum* (On the Philosopher's Stone). This book was a great success and a second edition called *Novum lumen chymicum* (New Light on Alchemy) came out the same year. Sendivogius's principal claim to fame was the story, universally believed, that, in 1604, he performed, in the presence of the Emperor Rudolph II of Poland, a transmutation of a base metal into gold. This event is celebrated in a magnificent painting by Wenceslas Brozik, which is in the Fisher Collection, Pittsburgh, USA. There is also a fine painting by Jan Matejko showing Sendivogius at work producing gold for King Zygmunt III of Poland.

Any attempt to read the works of the fifteenth century alchemists, such as those of Sendivogius, will give you an idea of what the unfortunate Newton was up against. Sendivogius and other alchemists seemed to be obsessed with secrecy. It was thus their practice to express everything in coded language, known only among themselves. In his *Treatise on Salt*, Sendivogius writes: '. . . the best of all the Stones is found in the new Habitation of Saturn, which has never been touched; that is to say, of him whose Son presents himself, not without great Mystery, to the Eyes of all the World, Day and Night . . .', and so on. One is tempted to suspect that alchemical secrecy had a motive other than the preservation to the elect of worthwhile ideas. It might equally serve to conceal the essential fatuity and futility of the whole enterprise.

Although Newton devoted as much, if not more, effort and time to alchemy as he devoted to real science, it is hardly necessary to report that nothing came of this labour except a mass of spoiled paper. Some people, comparing this disappointing outcome with the magnificent achievements of his scientific work, might conclude that the reason was that alchemy was, and is, a farrago of wishful-thinking nonsense. Clearly Newton did not think so.

> Ultraviolet rays exert a tonic influence on our health.
> *Sir J. Arthur Thomson (1890–1977), Professor*
> *of Natural History, University of Aberdeen*

A martyr to chemistry

Karen E. Wetterhahn, an attractive woman of 48 and a talented research chemist, was professor of chemistry at Dartmouth College in southeastern Massachusetts. Professor Wetterhahn was studying the way in which heavy metal poisons exert their harmful effect. She was engaged in a study, using nuclear magnetic resonance (NMR) spectroscopy, of how mercury binds to a protein associated with DNA. The idea was to find out what part of the protein molecule the mercury ion attached to, and this could be done by NMR. Other metals such as copper and zinc do not give NMR signals.

In order to get the best possible signal NMR signal Professor Wetterhahn used the mercury compound dimethylmercury. In August 1996 she was pouring some of this colourless liquid compound into an NMR tube when she accidentally spilled a tiny amount of it onto her latex glove. Within 15 seconds this volatile and fat-soluble fluid passed through the glove and was absorbed into the skin of her hand. A few months later, Professor Wetterhahn developed symptoms of severe mercury poisoning and, in less than a year from the time of the accident, she died from this cause.

Distressed colleagues called for publicity of the fact that latex and PVC gloves offered no protection against this highly toxic compound, and recommended that a much safer mercury compound be substituted.

Biology

> I mistrust laboratory methods because what happens in a laboratory is contrived and dictated. The evidence is manufactured: the cases are what reporters call frame-ups. If the evidence is unexpected or unaccountable it is re-manufactured until it proves what the laboratory controller wants it to prove.
>
> *George Bernard Shaw (1856–1950)*

Spermatic homunculi

When the Dutch amateur scientist Anton van Leeuwenhoek (1632–1723) invented a single-lens microscope and became famous for his researches into the microcosm – reported to the London Royal Society from 1673 onwards – he had many imitators. Some of these were more noted for their imagination than for the accuracy of their observations, but, perhaps, they can be excused on the grounds of the poor quality of the lenses they were using. Leeuwenhoek claimed to have been the first to describe spermatozoa, in 1677, and, although this claim is disputed, there is no doubt that he minutely described sperms and did so with remarkable precision.

The findings of some of his successors were fairly obviously conditioned by their preconceptions. It is hard, otherwise to account for the fact that several of them observed tiny men or women inside the heads of sperms and made drawings to prove it. To these observers it was obvious that all their little homunculi needed was a comfortable womb in which to grow. This finding also explained the previously inexplicable fact that some sperm produced boys and some girls.

These were not the only people whose ideas conditioned their observations – or the record of their 'findings'. Another remarkable example is described in the next section.

√

> All species were created separately. No new species have occurred since the Creation.
> *Carolus Linnaeus (1707–78), the inventor of the binomial (genus/species) nomenclature for the naming of plants*

Wishful thinking of a biologist

The German biologist Ernst Haeckel (1834–1919) was professor of comparative anatomy at the Institute of Zoology in Jena. He was the first German biologist to provide unequivocal support for Charles Darwin's theory of evolution. In fact, so enthusiastic was Haeckel that he discovered proofs of the theory that actually did not exist.

In the course of development, the human embryo passes through stages in which it clearly resembles, successively, the embryos of a range of organisms more primitive than ourselves. It is as if each of us repeats in our bodily development the evolutionary history of humans. This observation seems to have been made first by the German embryologist Karl Ernst von Baer (1792–1876) who published a two-volume textbook of embryology in 1828 and 1837. In this he pointed out that vertebrate animals which were quite different in their fully developed form had embryos that were remarkably similar in their early stages.

Haeckel was greatly taken with the idea that humans went through these stages as embryos, and came up with a phrase that, for many years, echoed though the corridors of the biological and natural history scientific institutions. The phrase was: 'ontogeny recapitulates phylogeny'. Ontogeny is the developmental history of an individual from the time of conception, and phylogeny is the sequence of events involved in the evolution of a species.

Phrases like this can be dangerous. They sound so impressive and convincing that people are apt to take them uncritically on trust. But Haeckel did not depend on a single phrase. He published remarkable drawings to support his claims. These drawings compared the embryos of species such as pigs, oxen, rabbits, fish and humans and showed in a remarkably convincing way that, throughout the embryonic stages, you would be hard put to say which was which. Haeckel went further and produced drawings to support his contention that the more closely two animals resemble one another in bodily structure, the longer their embryos remain indistinguishable. On the basis of this and similar evidence he concluded that we had 'sufficient definite indications of our close genetic relationship with the primates'.

As is often the case with scientists, Haeckel was a little carried away by his observations and was inclined to make more of his theory than was, perhaps, justified by the facts. If the embryos of different classes of animals, such as birds, reptiles or mammals, are compared at certain stages, it is often very difficult, by inspection, to say which is which. But Haeckel's claim that, in their development, the 'higher animals' actually pass, in turn, through all the stages shown by simpler animals is simply not true.

There is a reason why the embryos of a wide range of vertebrate animals resemble each other. It is simply that the structure of the human body is much closer to that of other species than most people realize. When the anatomies of the vertebrates are compared, it is seen that they differ essentially only in the shape and size of individual parts rather than in possessing different parts. The similarities between even remote vertebrate species are far greater than the differences. Such differences as distinguish them in their adult form appear only in the course of relatively late development.

Many scientists contributed to this understanding, perhaps the most outstanding being the English comparative anatomist Richard Owen (1804–92), who showed the equivalence of many anatomical structures not only in terms of their position in the body but also in their developmental origins. Owen's concept of homology was illustrated by examples of organs in different species that not only shared the same essential anatomical structure (if not necessarily the same function) but also developed from the same germinal layer in the embryo.

How did Haeckel get away with his theory for so long? It was largely, unfortunately, on the basis of his drawings. These became famous and have been repeatedly reproduced by publishers over the course of the last 120 years or so. But in 1997 someone called Haeckel's bluff. The embryologist Michael Richardson, writing in the learned journal *Anatomy and Embryology*, describes how he and his associates made a direct comparison of the features of 50 vertebrate embryos with Haeckel's drawings. They discovered that Haeckel had cheated. Some features that did not fit in with his theory had been omitted from the drawings, and some that helped to promote it had even been added.

This was not the first time that Haeckel had been challenged. When, during his lifetime, he was asked to explain certain discrepancies, he stated that he had made the drawings from memory. This was not considered a good enough defence and he was convicted of fraud. The really extraordinary thing, as Michael Richardson recorded, was that, in spite of this, Haeckel's drawings continued to be used.

While Haeckel's theory of recapitulation was, for a time, almost universally believed, the debunking of Haeckel was not widely known. Haeckel lived at a time when rather woolly pseudo-philosophical ideas were immensely popular. Haeckel taught, for instance, that cells had souls. His book *The Riddle of the Universe*, an extraordinary mishmash of real science and imaginary nonsense, was a great popular success and ran into numerous editions. So, although his ontogeny ideas were brushed aside at a fairly early stage by the serious scientists, they continued to be accepted by the lay public.

It is hard to decide whether Haeckel was a deliberate fraud. The probability is that he was not. When people have a strong proprietary interest in a theory, there is a real tendency for selective judgement and even, perhaps, for selective observation. This is the mechanism by which so much pseudoscience is perpetuated. Almost unconsciously, you latch on to what supports your theory and ignore uncomfortable facts that do not. We all do it, but scientists are supposed not to.

> I enclose a copy of my paper on inheritance of characteristics in plants.
>
> *Gregor Mendel, writing to Professor Karl Wilhelm von Nägeli (1817–91), the ablest botanist of the time, who did not reply, and Mendel died without any recognition that he had founded the science of genetics*

The protein gene blunder

At the end of the nineteenth century, chemistry had advanced to the stage at which quite a lot was known about the structure of proteins. Their 'building blocks', the amino acids, were known and it was clear that, with 20 amino acids – any of which could be linked together in any order – the number of permutations for proteins was enormous. It was, most experts thought, the only chemical molecule with the necessary complexity to be the vehicle of inheritance.

By the first few years of the twentieth century, it had become clear that inherited characteristics were transmitted in the chromosomes of the sperm and eggs. With only 46 chromosomes in each human cell, and thousands of characteristics, each one must carry a lot of information. So, in 1909, the Danish botanist Wilhelm Johannsen (1857–1927) decided that each chromosome must carry many separate units, each one being responsible for a single feature. Johannsen called these units 'genes'. The term was taken from the more general term 'genetics'.

But what was the real nature of these genes? By the 1930s it was known that chromosomes contained two types of sugar molecule, a phosphoric acid molecule, four small basic (alkaline) molecules (two called purines and two called pyrimidines) and proteins. Of these, it was believed that only the proteins would be any good at carrying many different items of information. Apart from the protein there was room for only seven characteristics. So everyone concentrated on the protein and regarded the other constituents of the chromosomes as vaguely interesting but irrelevant. These other constituents were linked together to form a large molecule that was called 'nucleic acid' because most of it was found in the nuclei of cells.

For over 40 years this was how things stood: only proteins could be diverse enough to carry genetic information. Then, in 1944, an extraordinary discovery was made. Some bacteriologists were studying strains of the bacterium that causes lobar pneumonia – the pneumococcus. There were two strains of the organism, one with a smooth capsule and one without a capsule. One could build itself a capsule; the other could not. This was obviously a genetic difference. So the scientists mashed up some of the capsulated bacteria, made an extract from them, and added the extract to the strain without a capsule. This converted the non-capsulated type into the capsulated type. What is more, all the subsequent offspring of the converted bacteria had capsules. Which component of the extract was carrying the instructions for making the capsule? When the extract was carefully analysed, it was found that it contained no protein at all. The only thing there was nucleic acid – sugars, phosphoric acid, and purine and pyrimidine bases.

At first the scientists thought there must have been some mistake, but the same results were obtained when the experiment was repeated: only the nucleic acid could transform a capsule-less bacterium into one with a capsule. The stunned scientists were forced to admit that they had been wrong for years: the neglected nucleic acid was the carrier of the genetic information, not the protein. Less than 10 years later, Watson and Crick showed them how it was done.

In the people called Eugenists, in the faith that they hold, lies the hope of the world. Every man is part of the world; it is for every man to see that his part of the world is as good as it can be. That is the foundation of the youngest and greatest of all the sciences – Eugenics.

Arthur Mee (1875–1943), English writer and editor,
writing in the introduction to Harmsworth's
seven-volume encyclopedia Popular Science

Francis Galton's great misconception

By any standards Francis Galton was a successful man. A cousin of the great pioneer of evolution, Charles Darwin, he was born in 1822 into a happy devoted family. He was destined for the medical profession and, while still a teenager, made a tour of medical institutions in Europe. This gave him a taste for travel and he spent months wandering about the Middle East before settling down to study in Cambridge. There, he worked so hard that he had a breakdown and left without taking his degree.

Fortunately, his talents soon overcame this minor disadvantage and before long he had been appointed a Fellow of the Royal Geographical Society and, the highest scientific distinction of all, a Fellow of the Royal Society, at the age of only 34. His interests were wide and he rapidly assimilated almost all that was known in a number of different disciplines. He wrote a classic handbook called *The Art of Travel*; he studied colour-blindness to such effect that, for many years, it was known as 'Galtonism'; he researched the physiology of visual memory and of the senses of taste and smell; he mastered meteorology and invented the word 'anticyclone' to describe the phenomenon he had discovered; he developed the branch of statistics known as correlational calculus; he studied twins to determine the relative effects of nature and nurture; he developed fingerprints as a means of criminal identification and persuaded the police authorities at Scotland Yard to adopt his system; he was interested in blood transfusion, and spent years developing improved standards of measurement; he even carried out a statistical trial of the power of prayer and caused a religious furore when he published the negative result. Altogether he wrote nearly 200 scientific papers and nine books.

As a child, Galton had been exposed to conventional religious teaching in school and church, and was deeply disturbed at his inability to reconcile the literal statements of the Bible about Creation with the observed facts of science. So Darwin's *Origin of Species* came as a profound relief and allowed him to clear his mind of what he had come to regard as a mass of superstition.

Galton was a life-long enthusiast for measurement and, inevitably, turned his attention to anthropometry, the measurement of human qualities, both physical and mental. This became an absorbing interest. Associated with this interest was an intense concern with heredity, and these preoccupations led to the publication of books such as *Hereditary Genius, Inquiries into Human Faculty and Its Development, English Men of Science: Their Nature and Nurture, Noteworthy Families,* and *Natural Inheritance.* These books make fascinating reading and throw as much light on Galton's beliefs and prejudices as they do on their subjects.

Hereditary Genius is concerned with establishing the laws of inheritance and the consequences of Galton's findings. Charles Darwin loved the book. As he said to Galton: 'I do not think I ever in all my life read anything more interesting and original.'

Modern critics, while largely agreeing with Darwin, might be inclined to point out a few flaws. Galton's enthusiasms often seemed to run away with him. In this book he makes the case for the hereditary nature of many mental and physical characteristics that predispose people to various professions and activities. In effect, he was saying that heredity determines the adoption of such jobs as high court judge, statesman, military commander, novelist, poet, scientist, painter, churchman, classical scholar, oarsman, wrestler and criminal. The labours he expended compiling the great mass of data he presents on these categories must have been enormous.

This work persuaded Galton that it was the duty of science to ensure that the best of these characteristics should be passed on to future generations, thereby improving the stock of humankind. Since all good ideas demand a good title, Galton came up with the term 'eugenics', which means 'well-born'. It must be remembered that Galton sincerely believed that moral qualities were inherited. So any scheme that increased the stock of people of high moral character must be of enormous benefit to society. Galton's ideas of superiority were, of course, strongly conditioned by the values of his upper middle class upbringing. His father had been a banker and a cultivated and intellectual man. It never occurred to people in Galton's position, especially at that stage in social evolution, that there could be any alternative criterion of status. Such people just naturally thought in terms of overall superiority and inferiority. Society was strongly stratified and most people knew their place. The problem was how to set about implementing eugenics.

Galton knew, of course, that for centuries livestock farmers had been altering the qualities of their animals by selective breeding. The ancient Greeks held that men distinguished in war or in other valued activities should be encouraged to breed, while people seen as socially undesirable should not. This can be achieved in two ways: by active encouragement of mating to promote desirable physical or mental qualities, or by discouragement of mating to eliminate undesirable qualities. The former is called positive eugenics, and the latter negative eugenics.

The first step in any eugenics programme is to identify good and bad qualities. This is a matter on which there can be no firm consensus of opinion. The next step is to decide whether such qualities are genetically determined – a point which begs the entire question. Galton's notion, for instance, that morality was a genetic entity is complete nonsense. Morality is an arbitrary entity determined by the social mores of the time and has no absolute features. The real danger of eugenics is quite different and it is this that makes Galton's efforts a scientific blunder of the first magnitude. This is best illustrated by referring to what actually happened when, in the early part of the twentieth century, eugenics was generally regarded as the best hope for mankind and was almost universally approved of by the Establishment.

Galton was disturbed by two awkward facts: modern civilized living was protecting society against Darwinian natural selection so that inferiors were able to survive, and, worse still, the 'best' people had few children while the 'inferior' classes bred like rabbits. This was a distinctly anti-eugenic phenomenon and was likely to lead to a steady degradation in the overall quality of humanity. Something had to be done about it. Obviously either the top people had to be persuaded to do their duty more enthusiastically or the bottom people had to be persuaded to do it less. Galton was realistic enough to appreciate that few people would choose to have more children than was convenient simply out of a sense of social responsibility. So he advocated governmental action to provide special allowances to encourage higher breeding. For 40 years Galton strove to promote eugenics. The topic became highly respectable. Scientific textbooks of the period referred to it approvingly. Popular science books were full of enthusiasm for eugenics and contained chapter headings such as: 'Nations of the future', 'The new mankind', 'Eugenics and the family', 'From Galton to Ellen Key', 'Eugenics through love', 'The Eugenics Congress', 'Preventive eugenics', 'National eugenics', and 'Eugenics and the future'. For a time, eugenics almost seemed like a religion. Even painters were inspired to depict the marriages of ideal mates. Galton's efforts were strongly approved of by Government and earned him a knighthood. Needless to say, all this talk and writing on positive eugenics came to nothing.

But that was by no means the end of the matter. On the principle that power corrupts, what did happen was that 'well-meaning' eugenicists were strongly advocating the sterilization of classes of people thought to be of defective stock – people with psychiatric disorders, epilepsy, learning disadvantages, or a history of criminal activity or sexual 'deviation'. Astonishingly, several European countries and no fewer than 27 states in America enacted laws providing for the voluntary or compulsory sterilization of such people.

In Sweden, for instance, thousands of women were sterilized without their consent on eugenic grounds. Altogether, some 60,000 people were sterilized under a 1926 Swedish law, on the grounds that they had 'undesirable' characteristics including 'an unhealthy sexual appetite'. Some were even sterilized because of poor eyesight. The culmination of this darker side of eugenics was, of course, Adolf Hitler's attempt to produce a 'master race' by encouraging mating between pure 'Aryans' and by the murder of six million people whom he claimed to have inferior genes.

It is hardly fair to Galton to blame him for the Holocaust or even for his failure to anticipate the consequences of his advocacy of the matter. But he was certainly the principal architect of eugenics, and

Hitler was certainly obsessed with the idea. So, in terms of its consequences, this must qualify as one of the great scientific blunders of all time.

Eugenics is still with us. One unperceived example of positive eugenics is the selection of sperm donors for artificial insemination of women whose husbands are sterile. Doctors who perform this procedure try to select donors, usually from the ranks of medical students, who are mentally and physically well above average. Such donors are, of course, anonymous. The Repository for Germinal Choice, a French institution founded in 1979, maintains stores of sperm from a number of Nobel Prizewinners for the use of women who desire it. The Singapore government followed Galton exactly when it adopted a positive eugenics measure in 1983 by legislating for the provision of educational advantages for children born to educated women.

Clearly, eugenics isn't going to go away, and we had better be on our guard.

> War is a relic of barbarism probably destined to become as obsolete as duelling.
>
> *Lord Kelvin (1824–1907)*

The Lysenko blunder

In spite of the almost universal recognition, by the beginning of the twentieth century, that all animals, including man, acquire their hereditable characteristics at birth, the Russian agronomist Trofim Denisovich Lysenko (1889–1976) had come to believe in an alternative theory of evolution. This he had derived from an uneducated plant breeder who had put it forward as an explanation for the hybrid plants he had developed. The theory – a revival of a long-discredited idea – was, essentially, that characteristics acquired during life could be passed on to the offspring (see The great Lamarck blunder). Working in the All-Union Genetics Institute at Odessa in the early 1930s at a time when there was a crisis in Russian agriculture, Lysenko claimed that he could alter the genetic constitution of strains of wheat by exposing them to various environmental factors. His experiments were crude and their results ambiguous, but Lysenko promised, by his new methods, to provide a great increases in crop yields. Although to the rest of the scientific world this was nonsense, it was welcomed by the authorities, and Lysenko became a hero. Not only did Lysenko's ideas seem to Stalin to fit in with Communist thinking, but they would also solve the most pressing problem of the time: if wheat could be genetically improved by changing its environment, then it might be possible to improve humans, given the correct political, social and economic system.

From the time of the summer meeting in 1948 of the Academy of Agricultural Science, it was known that Stalin endorsed Lysenko's theory. From then on, any belief in conventional genetics was interpreted as disloyalty to the State. A number of scientists had bravely protested. These included Nikolai Vavilov, the Director of the Institute of Genetics of the USSR Academy of Sciences. Vavilov had studied genetics at the University of Cambridge and, on his return to Russia, had established 400 research institutes throughout the country. He had travelled widely and had amassed an enormous collection of plants including over 30,000 specimens of wheat. After Lysenko's ideas had been proclaimed official, Vavilov's days were numbered. Between 1934 and 1939 Lysenko publicly denounced Vavilov at successive plant-breeding congresses. In 1940 he was arrested and sent to a concentration camp at Saratov. Lysenko was appointed in his place and remained the leading force in Soviet biology until well into the middle 1960s, thereby setting back Soviet genetics research and teaching for many years.

In 1953 the nuclear physicist Andrei Sakharov (1921–89) became the youngest member ever to be appointed to the Soviet Academy of Sciences. At the 1964 elections for the membership of the Academy, one of Lysenko's closest associates and supporters, Nikolai Nuzhdin, was up for election to full membership. When Sakharov heard this, he decided to make a stand. Remembering the 'tragedy of Soviet genetics and its martyrs' as he put it, he opposed Nuzhdin's election, saying: 'The charter of the Academy requires that its members should be scientists of the first rank as well as people of high civic responsibility. Nikolai Nuzhdin is neither and does not satisfy the requirements. He and Academician Lysenko are responsible for the shameful state of Soviet genetics and biology, for promoting pseudoscience, for degrading scholarship, and for defaming, ruining, and bringing about the arrest and even the death of many real scientists. I urge you to vote against Nuzhdin.'

Premier Khrushchev was furious when he heard what Sakharov had said. He ordered the Chairman of the KGB to start accumulating compromising evidence against Sakharov. Soon after this, Sakharov began to become estranged from the authorities. His persistence in campaigning for a nuclear test-ban treaty and for civil rights in the USSR exacerbated the situation. That he was awarded the Nobel Peace Prize in 1975 only made matters worse. This was interpreted as Western propaganda. In 1980 he was sent into exile in the closed city of Gorky. It was not until Gorbachev gained power that he was rehabilitated.

This blunder and odd sidelight in the history of science had no effect on world-wide scientific opinion. For, long before Lysenko's time, all free-thinking scientists were well aware that acquired characteristics could not be genetically transmitted. Multicellular organisms are formed by repeated division of the fertilized egg cell with exact replication of the DNA. So every daughter cell contains identical copies of the DNA of that original fertilized cell. This DNA is the determinant of the physical characteristics – the phenotype – of the organism. During its lifetime, an organism can undergo all kinds of changes as a result of environmental effects. It can suffer damage or it may enjoy physical advantage. But these are changes in the body, not in the DNA of the sex cells, i.e. they are changes in the phenotype, not in the genotype. So the genes that an organism passes on to its offspring are those it acquired at the moment it came into existence. Modifications in species come about by natural selection, as explained by Charles Darwin, not by things that happen during the lifetime of the members of species.

Darwin's theory was presented to the world in 1858. Mendel established the principles of inheritance in the 1870s. All of this Lysenko brushed aside. Darwin's principal supporter was Thomas Henry Huxley (1825–95). Huxley's grandson, the biologist Julian Huxley (1887–1975), was at his peak during the Lysenko fiasco. A remark worthy of his illustrious and witty grandfather is attributed to him as a comment on the official Soviet doctrine of the inheritance of acquired characteristics: 'If this theory is correct, it would follow that all Jewish boys would be born without foreskins.'

In 1953 Watson and Crick established the chemical structure of the DNA molecule and provided an explanation of how genes make perfect copies of themselves, but Lysenko was unmoved. It is hardly necessary to say that Lysenko's ideas were as disastrous to Soviet agriculture as they were to Soviet biology. By the time it was appreciated that his claims were not going to be fulfilled, however, he and his cronies were well established politically. Even after he lost his job as director of the Institute of Genetics in 1965 – the year after Khrushchev was deposed – Lysenko's followers retained their academic positions, titles and degrees, and continued to exert their influence on Soviet biology.

The moral, of course, is that when political ideology is allowed to influence science, it is all too liable to prove to be the enemy of truth. The same applies to religious dogma.

> . . . it is only the other day that an eminent physiologist, Dr Brown Sequard, communicated to the Royal Society his discovery that epilepsy, artificially produced in guinea pigs, by a means which he has discovered, is transmitted to their offspring.
>
> *Thomas Henry Huxley (1825–95), the principal exponent of Darwinian evolution, apparently accepting the possibility of the inheritance of acquired characteristics*

The wrong dogma

In the exuberance of intellectual excitement that followed the discovery of the structure of DNA and the genetic code, a solemn pronouncement was made. This was so solemn that it took on a title that should certainly be rare in science: it was called the 'central dogma'. Now dogmas sit uncomfortably in science, as scientists do not like being told what to believe. For a time, biological scientists were either a bit unhappy about this name or thought it rather a joke. Nevertheless, the phrase caught on and appeared in many textbooks.

So what is the 'central dogma'? In the late 1950s, Francis Crick (1916–) put forward the hypothesis that there was only one direction in which it was possible for the transcription from the genetic code to protein to go. The direction was DNA to RNA to protein. This was how things were and there was no way it could be otherwise. Protein could not give rise to RNA and RNA could not give rise to DNA. It seems that this hypothesis was considered so strong – almost self-evident – that, soon afterwards, Crick coined the phrase the 'central dogma'.

Apart from the word 'dogma' no one was arguing. The genetic code is a sequence of chemical bases that, with a double backbone of sugars and phosphate, form the DNA molecule. A gene is a length of DNA and the sequence of bases, taken three at a time, specify the identity and sequence of amino acids for which the gene forms the code. Amino acids link together to form proteins. Proteins are not formed directly from DNA; there has to be at least one intermediate molecule involved in the transfer of this information from the DNA to the site of protein synthesis. This intermediate is called messenger RNA and it carries the code of the gene to the ribosomes where protein is assembled.

DNA is a double helix, often likened to a ladder with the 'rungs' made up of paired bases. This structure is only possible if the bases link up in a specific way: guanine pairs with cytosine and adenine pairs with thymine. Each base is said to be complementary to the base with which it pairs. When part of a DNA molecule needs to be 'decoded' to produce a protein, the two strands separate and are used as templates to make RNA molecules of complementary sequence. These RNA molecules move out of the nucleus to the cytoplasm, where small structures called ribosomes attach to them.

In the ribosomes the base sequence is 'read' three bases at a time, and a chain of amino acids is assembled, based on the 'instructions' in the RNA, to make a protein. Thus DNA makes RNA makes protein. But is the 'central dogma' true? Is this just a one-way process?

It is now known that RNA to DNA transfer is possible. In 1983 a researcher in the Pasteur Institute, called Françoise Barré-Sinoussi, isolated a virus from an AIDS patient, the virus that is now called HIV. This proved to be a rather unusual virus although not the first of its kind to be found. It was an RNA virus, i.e. its genetic material was RNA not DNA, and it had an enzyme with the remarkable property of allowing the virus's RNA to be used as a template to form a DNA copy, which the virus then inserted into the DNA of the host cell it had invaded. Because of its extraordinary property of going in the wrong direction in the dogma sequence, this enzyme is called reverse transcriptase. It catalyses transcription in the reverse of the usual direction, from RNA to DNA. So the virus is called a retrovirus.

This was all very embarrassing for the dogmatists. No one likes to admit that he or she has been totally wrong about something. Interestingly, some of the latest textbooks still refer to the central dogma. One even states that the central dogma 'defines the paradigm of molecular biology' and goes on to state that the paradigm is 'that genes are perpetuated as sequences of nucleic acid, but function as being expressed in the form of proteins'.

This is a neat way of getting round the problem by restating its terms. The same excellent work has the grace to pay lip service to the original statement of the dogma by admitting that 'the restriction to unidirectional transfer from DNA to RNA is not absolute. It is overcome by the retroviruses . . .'

> I know of a professor of physiology who, on being
> invited to view an experiment on the images of the
> eye, retorted, with annoyance: 'A physiologist has
> nothing to do with experiments.'
>
> *Hermann von Helmholtz (1821–94),*
> *the great German physiologist and physicist*

The SV40 near miss

This is an account of a blunder that might well have become a disaster but was stopped in the nick of time.

SV40, short for simian virus number 40, was the fortieth virus to be detected in the simian (monkey) family of viruses. It is of special interest because it is known to be able to transform normal cells to cancer cells. For this reason it is known as an oncovirus (*onco* is Greek for a lump) and it acts by virtue of certain genes it carries called oncogenes.

One of the most important pioneers of genetic engineering was the American molecular biologist and Nobel Prizewinner Paul Berg (1926–). It was he who first discovered transfer RNA, the type that is involved in the putting together of a chain of amino acids to synthesize proteins. Each molecule of transfer RNA is specific for a particular amino acid and carries a code in the form of a triplet of bases that is complementary to a triplet on messenger RNA. Berg's Nobel Prize, however, was awarded for the important part he played in developing the DNA splicing methods that made recombinant DNA techniques (genetic engineering) possible.

Berg was studying viruses that caused cancer in animals – an obviously important line of research. Viruses can enter human cells and take over some of the cells' functions. Since cancer was thought to have something to do with changes in DNA, viruses offered an ideal way in which possible cancer-causing genes could be introduced into human cells. Berg's research was targeted on finding out what viruses did when they entered cells and how they affected the normal working of the cells. He decided to use the monkey virus SV40.

The time was ripe. It had just been discovered that certain enzymes, called restriction enzymes, existed that could cut DNA at particular points to give fragments of DNA. These could then be inserted into the DNA of other cells where they would perform their normal function. Berg recognized that this provided him with a powerful new tool. If he were to cut up the DNA from an SV40 virus and insert the pieces, one by one, into an animal cell, he should be able to isolate the genes that caused cancer – a highly exciting prospect.

At this stage, Berg suggested to one of his research assistants, Janet Mertz, that she should insert some of the SV40 DNA fragments into the common bacterium *Escherichia coli (E. coli)* to see whether any of the virus's genes were carcinogenic. *E. coli* is an organism that all of us carry in countless millions inside our intestines and it is extensively used in genetic engineering. Most strains of *E. coli* are fairly harmless; a few are quite dangerous, but these strains are rarely encountered. By the middle of 1971 Janet Mertz was ready to start doing what Berg had proposed. At this stage, however, she attended a genetic engineering conference and happened to mention what she was about to do. Her colleagues were horrified. They included Robert Pollack, a cancer expert working at the Cold Spring Harbor Laboratory under James Watson of double helix fame. 'If this escapes from the lab,' he said, 'you will have SV40 replicating in step with *E. coli.*' The thought of a cancer-causing *E. coli* at large in the world and spreading like wildfire throughout the human population did not bear thinking about.

No one was more horrified than Berg at what he had nearly done. The experiment was immediately stopped and Berg began to pay a lot of attention to the dangers and ethics of this kind of work. In July 1974 Berg, with the full approval and backing of many of the leading researchers in the field, published a famous letter in *Science*. This gave clear warning of the dangers inherent in recombinant DNA technology and proposed a voluntary moratorium on the most obviously dangerous experiments and strict safety precautions for the control of others.

In 1976, following an international conference on safety in genetic engineering held in Asilomar, California, official guidelines were issued by the American National Institutes of Health.

Medical Science

X-rays will prove to be a hoax.

*Lord Kelvin (1824–1907), who had been appointed
a professor of mathematics and physics at the age of
22, and who became the greatest scientist of his day*

Magnets for toothache

Sometime around 500 AD there lived a Greek physician called Aetios or, in Latin, Aetius. Very little is known about him other than that he wrote a medical textbook. In this book, one of the cures he recommended was the use of the lodestone – the naturally magnetic mineral magnetite that was used by early mariners as a crude directional compass. Aetius says: 'We are assured that people troubled with gout in their feet or hands obtain relief simply by holding a lodestone.' This is probably the earliest known reference to this form of treatment.

The recommendation was taken up by the notorious Paracelsus (*see* The opium blunder) and, from then on, as has often been the case in medicine, one 'authority' copied it from another's writings and so the idea was perpetuated through the ages, acquiring, with each new reference, additional proof of its correctness. Paracelsus insisted that the magnet was effective in bloody fluxes and other bleedings as well as in many diseases. The fifteenth-century physician Marcellus assured us that the magnet would cure toothache, as did the sixteenth-century physician Leonard Camillus. J. J. Wecker, in his book *De secretis*, claims that, applied to the head, the magnet will cure headache; Kircher prescribed lodestone, to be worn about the neck, as a preventive of epilepsy; and at the end of the seventeenth century, doctors were selling magnetic toothpicks for toothache and magnetic earpicks for earache. As late as 1846, in his *History of Inventions, Discoveries and Origins*, Professor John Beckmann of Göttingen refers to the 'fashion in modern times for the use of the magnet to cure toothache'.

There is no known way in which the application of a magnet could have any effect whatsoever on gout, bleeding, headache, toothache or any other disease or symptom. But doctors have always been able, by the display of an authoritative or learned manner, to make their patients believe any sort of nonsense. In addition, even among those members of the lay public with enough knowledge of science to dismiss claims such as this, there will always be some with enough latent superstition to give them the feeling that 'there might be something in it'.

> Of all the cases of St Vitus's dance that passed through my hands last year as physician to the Great Ormond Street Hospital, more than one third were clearly attributable to school causes. More than twice the number of cases were of girls than of boys, because of the smaller brain of the former and their more delicate organization. The causes of

> the disease are over-schooling, excitement at examinations, home lessons and caning.
>
> *Dr Octavius Sturges,*
> *writing in the* Lancet *of 15 January 1887*

The opium blunder

The species name of the common poppy, *Papaver somniferum,* is derived from the Latin word *somnis,* meaning sleep. No one knows when the remarkable properties of dried poppy juice were first discovered, but we can assume that this dates from a very early stage in the history of humankind.

There are Sumerian records of the use of opium dating back 7000 years. Early Assyrian tablets contain references to it. Homer described its use in the *Odyssey,* Hippocrates prescribed it, and Galen was enthusiastic about it. In the first century AD, the outstanding drug book was *De materia medica* by Pedanius Dioscorides (*c.*40–*c.*90). This book described opium and its properties. The practice of cultivating poppies to obtain opium seems to have started in Greece and Mesopotamia and then spread slowly eastward. So far as we know, opium was unknown in India or China until about the seventh century AD. Opium smoking did not start until after tobacco and smoking pipes were introduced to the Old World from America.

Raw opium is now harvested in many parts of the world, including India, Turkey, Iran, Russia, Bulgaria, China and Japan. It contains as many as 25 different alkaloids – nitrogen-containing alkaline plant products of complex chemical structure and having some drug action. Of these only morphine, codeine and papaverine are of significant medical value, and it is the morphine that has made opium so important, both medically and sociologically. Morphine is very difficult to synthesize, so, in spite of all the advances in pharmacological chemistry, it is still obtained from the humble poppy. It is the properties of morphine that make it important. Morphine is a highly effective painkiller and it remains the standard against which new painkilling drugs are judged. It produces a mental state of tranquillity in which distress of mind is relieved, and in some people it produces euphoria. The body quickly adapts to morphine and sets up a new physiological state that is dependent on the drug. When this has happened, withdrawal of morphine is extremely unpleasant.

Opium continues to be a widely used 'recreational' drug. Many people who have become accustomed to the effects of cocaine feel the need to move on to something with a more powerful effect on the psyche. Today, opium is still extensively smoked, even in Britain, where it is usually mixed with tobacco. Everyone is aware of the dangers of 'main-lining' heroin, but there is a widely held view that opium can be safely smoked without incurring the risk of addiction. This is nonsense. It is true that it will take longer for a true physiological addiction to occur with crude opium than with heroin (see below) but opium, if used repeatedly, will soon result in addiction.

Used illicitly, morphine is poisonous and has numerous undesirable side-effects. It depresses breathing, often fatally; it produces nausea and vomiting; constipation; dizziness; itching; urinary retention; low blood pressure; and mental clouding. The pupils of the eyes become narrowed to pin points but, as asphyxia supervenes, the pupils become wide. This occurs before death from morphine poisoning.

Opium was one of the few powerful and useful drugs to be used in the early days of medical practice. There were several preparations, but the best known was *laudanum*. This is a simple solution (tincture) of raw opium in alcohol. The dose was reckoned in drops. The term 'laudanum' was coined by one of the most remarkable rogues, quacks and self-advertisers in the history of medicine – Theophrastus Bombastus von Hohenheim (1493–1541), better known as Paracelsus. The latter version of his name he adopted to show that he was a better man than the first century Roman medical writer, Celsus. Paracelsus, for commercial reasons, kept quiet about the real constituents of his patent cure-all, but opium was the active ingredient of later preparations of laudanum, so this is probably what he used. He was a man who loved inventing new words – mainly to impress the simple-minded – and this one was a beauty. The word 'laudable' means 'worthy of praise'. Paracelsus charged a very high price for his remedy which, he insisted, contained powdered gold, powdered undrilled pearls and other expensive matter. There are written records of opium well before this date. In 1398, one writer recorded: 'Of popy comyth iuys that physycyens callyth Opium other Opion.'

Laudanum and other forms of opium were widely available in Europe from the sixteenth century onwards and were considered a sovereign remedy for all kinds of diseases and disorders until well into the twentieth century. The great English physician Thomas Sydenham (1624–89) strongly advocated laudanum and considered it more valuable than any other drug. 'Without opium . . .' he said, 'medicine would be helpless and crippled.'

The colourful near-quack 'quicksilver' doctor Thomas Dover (1660–1742) went further and recommended his celebrated Dover's Powder for nearly everything. This powder, the successor of which was still being freely prescribed in the mid-twentieth century, consisted of opium, saltpetre, vitriolated tartar, ipecacuanha and liquorice. The dose recommended by Dover was enormous. What Dover and other opium enthusiasts apparently failed to appreciate was the huge tolerance for the drug that people who took it regularly developed. Dover wrote: 'Some Apothecaries have desired their Patients to make their Wills and Settle their Affairs, before they venture upon so large a Dose as I have recommended. As monstrous as they may represent this, I can produce undeniable Proofs, where a Patient of mine has taken no less a Quantity than a Hundred grains, and yet has appeared abroad next day.'

Hugh Walpole, in a letter dated 1751, writes: 'Lady Stafford used to say to her sister, "Well, child, I have come without my wit to-day"; that is, she had not taken her opium.' Whether the early advocates of opium were aware of the addictive properties of the drug is not clear. The high probability is that they were and that they regarded this as a distinct commercial advantage, as do contemporary purveyors of the best-known derivative of opium.

The extent to which a person can become habituated to opium is remarkable. The normal pain-relieving dose is about one-third of a grain (20 milligrams). Cases are on record of people regularly taking doses of over 1000 grains a day, i.e. 2.28 ounces, or 3000 times the normal dose. Extreme cases are recorded. One man living in South Illinois is said to have consumed 2345 grains of opium every day. It has to be said, however, that opium is often cut with adulterants, so it is impossible to say how reliable such accounts might be.

Probably the most famous opium addict was Thomas De Quincey (1785–1859). At the age of 19, while still a student at Worcester College, Oxford, he took laudanum to relieve the pain of toothache. This was highly satisfactory, although he would have done better to see a good dentist. Within ten years, as he reported, he had become 'a regular and confirmed opium-eater'. It is not clear why he used this expression since his only source of the drug was the liquid, laudanum, of which he kept a decanter at his elbow. De Quincy described the steadily increasing dosage necessary to satisfy him. His life became ruled by the counting of drops. At his peak of consumption he was taking 8000 drops a day – a dose equivalent to over 300 grains of opium and many times the lethal dose for a person who was not habituated.

The only good thing that opium did for De Quincy was to provide him with a saleable subject for a book. He started a torrid affair with a prostitute in London after running away from home. After becoming addicted to opium and renting Wordsworth's former home, Dove Cottage at Grasmere, he got a girl, Margaret Simpson, pregnant. Later, at 32, he married her. A large family followed and De Quincy's financial situation went from bad to worse. He had aspirations as a writer and wished to become 'the intellectual benefactor of mankind.' For reasons that are unclear, but possibly because of the opium, he published almost nothing. At this point De Quincy decided to make a virtue of necessity and he wrote *Confessions of an English Opium Eater*. This appeared in the *London Magazine* in 1821, and was published as a book the following year. It was an immediate best-seller and made him famous. It is an odd book. Purporting to warn humanity of the dangers of opium, it actually presents a rather attractive picture of the pleasures of drug addiction.

De Quincy, encouraged by the reception of this book, continued to write. His *Lake Reminiscences* deeply offended Wordsworth and the other Lake poets and an attempt to revise the opium book under the title *Sighs from the Depths*, was a failure. The book was padded out with irrelevant autobiographical detail and the style was turgid and difficult. He became eccentric and solitary and used large doses of opium as a retreat from reality. In all he published more than 14 volumes. Apart from a few fine articles on literary criticism, notably *On the Knocking at the Gate in Macbeth*, only the original *Confessions* is read by anyone today.

One of De Quincy's closest friends was the poet Samuel Taylor

Coleridge (1772–1834). Coleridge was a far more talented writer and had a career of great distinction. It was his association with William Wordsworth that led to the production of the *Lyrical Ballads* and the development of a new, more realistic, form of poetry. His own poems include several masterpieces, notably *The Rime of the Ancient Mariner* and *Kubla Khan*. His *Biographia Literaria* was the greatest work of literary criticism of his day. Coleridge settled for a time with the Wordsworths at Grasmere. But he, too, was an opium addict. His favourite resource was called 'Kendal Black Drop' and was said to be four times the strength of standard laudanum. By 1802, as a result of his addiction, Coleridge had suffered a mental collapse that led to alienation from Wordsworth. Although he produced much excellent work after that, it was all concerned with criticism, theology and philosophical speculation.

In spite of considerable evidence of the dangers of opium, doctors continued to advise it and it was freely available, without prescription, from chemists' shops. Anyone could buy laudanum (tincture of opium), opium pills, paregoric or chlorodyne. Paregoric, or paregoric elixir, was camphorated tincture of opium flavoured with aniseed and benzoic acid. Another version, known as Scotch paregoric elixir, was the ammoniated tincture of opium.

The term 'chlorodyne' was made up from 'chloroform' and 'anodyne'. There were various preparations. One popular brand contained opium, chloroform, tincture of Indian hemp (marijuana), prussic acid, and other substances. Chlorodyne was immensely popular and was consumed by people of all social classes. The writer Ouida, in one of her novels, referred to a character 'Who could no more live without a crowd about her than she could sleep without chlorodyne'. The *Daily News* of 11 January 1887 referred to a person who was in the habit of taking enormous quantities of 'the patent medicine known as chlorodyne, which had the effect of stupefying her'. Dr Collis Browne's Chlorodyne was extensively advertised in newspapers, magazines and hoardings until about the middle of the 20th century.

All highly successful products have their imitators, and Dr Collis Browne's was no exception. Chlorodyne was also produced by a firm called Freeman and another called Towle. Children were not neglected. Dalby's Carminative Mixture for children contained opium, as did Godfrey's Children's Cordial.

'Women's problems' were not talked about, so women with severe period pains (dysmenorrhoea) had little resource other than patent medicines. This may account for the fact that, in the nineteenth century, the opium addiction rate among women was three times that among men. In the middle of the nineteenth century, the hypodermic syringe with its hollow needle was invented and opium or its derivatives could be given by injection. This gave rise to the extraordinary notion that, so long as opium was not taken by mouth, the 'hunger for it' (addiction) would not occur. The needle became especially popular as a means of mitigating the horrors of war for the soldiery. During the American Civil War about 400,000 soldiers became addicted to opium.

The early-nineteenth-century chemists' shops stocked many opium preparations. Powdered opium was mixed with soap and rubbed up into little pills that were then coated with chalk to stop them sticking together. Every chemist had his pill-rolling board. Plain opium pills were 'improved' by the addition of poisonous lead compounds. Opium was sold as powders, lozenges, plasters, liniments, 'confections' and even enemas. The most popular form was the solution in alcohol – wine of opium, more commonly known as laudanum. The famous Dover's Powders, 'used the world over', consisted of powdered opium and ipecacuanha. Conceivably, but improbably, Dr Dover's conscience led him to add an effective emetic to the dangerous opium.

Opium was sold freely under many different names and it is not at all clear whether purchasers were always aware of what they were taking. Some proprietary remedy names – Poppy-head Tea, Battley's Sedative Solution and Nepenthe – hinted strongly of their nature to the educated. But others certainly did not. The drug was, however, widely and readily available under its own name from many sources other than chemists' shops. In the middle of the nineteenth century, penny packets of opium were as readily available in British cities as they were in Chinese towns and villages. You could even, in some places, buy opium in solid one-pound blocks.

By the end of the nineteenth century serious drug-taking was a well-established feature of social life. At first it was largely confined to the 'criminal' classes, but soon opium usage extended into more respectable groups. The 'bright young things' of the early twentieth century were considerably interested in drugs and narcotics had become a world-wide problem. Opium was too mild for many, and the doctors and scientists tried hard to help the situation.

The *Lancet* of 3 December 1898 contains a short report: 'A new hypnotic, to which the name of 'heroin' has been given, has been tried in the medical clinic of Professor Gerhardt in Berlin. According to a communication made by Dr Strube to the *Berliner Klinische Wochenschrift* it is a product of the di-acetic ester of morphia, and it was discovered by Professor Dreser, chief of the chemical department of the Elberfeld Farben Fabriken.' Every new drug was, initially, pronounced enormously beneficial and, of course, entirely safe. Heroin – the name was derived from the 'heroic' state of mind said to be induced by the drug – was no exception. The drug was, initially, even credited with the power of curing morphine addiction. There is no saying how many patients became thoroughly confirmed in their addiction by this bizarre form of treatment. Heroin is much more readily absorbed than morphine, whether taken by mouth or by injection, but, <u>once absorbed, heroin is rapidly converted to morphine</u>. The increased solubility, however, makes it easier for heroin to get quickly to the brain than in the case of morphine.

Because of its rapidity of absorption, and hence effect, heroin is more liable to produce addiction than plain morphine. In consequence it has been banned for medical use in almost all countries of the developed world except Britain. British doctors have insisted that the increased solubility of heroin justifies its use in the control of severe pain in terminally ill patients.

So how is it that we have got into such a mess with opium? Why was opium such a scientific blunder? The blunder was twofold: the first was the failure to protest against the cynical exploitation of opium for gain and to warn a gullible public about the dangers of the drug; the second, and more fundamental, scientific blunder was the failure to recognize that many, possibly most, humans are motivated almost exclusively by the desire for sensual gratification. It did not require a high level of scientific sophistication to reach this conclusion; and it was certainly possible to see this long before the current plague of narcotic usage was allowed to get under way. The desire for gratification manifests itself in many ways – overeating, alcohol indulgence, sexual promiscuity, purveyed entertainment and the inability to resist anything that promises any kind of pleasure. So much is self-evident. The blunder was to fail to see that, people being what they are, widespread 'recreational' morphine usage was a disaster waiting to happen. All that was needed to make it happen was the growth of affluence and the loosening of social restraints. The greed and moral bankruptcy of the purveyors could be relied on to do the rest.

> Listerian antisepsis is absurd in theory and impossible in practice.
>
> *George Bernard Shaw (1856–1950)*

The Charcot shaker and the Tourette trembler

The French physician and neurologist Gilles de la Tourette is remembered for his description of the extraordinary syndrome that now bears his name. This syndrome initially involves ever-worsening grimaces, tics and twitches and then progresses to its most distressing feature, the compulsive need to utter obscene words or remarks. This feature, known as coprolalia, occurs in about half the cases and in these the condition becomes a severe social disability. Tourette had no idea of the cause of the syndrome and, even today, this remains obscure. The earlier ideas, largely due to Freud, that it was the result of unresolved sexual or other conflicts have now been abandoned.

Tourette, like Sigmund Freud, was a pupil of the great Dr Charcot (1825–93) at the Salpêtrière in Paris. In 1892 Charcot made a momentous discovery – that patients suffering from Parkinson's disease, then known as paralysis agitans or the shaking palsy, were better after long journeys by carriage or rail. The rougher the journey, it seemed, the more these unfortunates benefited from it. Charcot immediately had a special chair made, based on the idea of the mechanical sifter, that was caused to oscillate vigorously from side to side by an electric motor and a cam mechanism.

Dr Gilles de la Tourette was deeply impressed by Charcot's oscillating chair and watched with fascination as Charcot's patients tolerated with seeming approval a 15-minute ordeal that, to a normal person, would be extremely unpleasant. Afterwards the patients seemed to be remarkable improved. They were less fatigued, able to stretch their limbs and slept well at night. This, it seemed to Tourette, was a whole new modality of treatment, and one that could be applied to his own patients. Wasting no time, he arranged for a vibrating helmet to be constructed. This, too, used an electric motor, turning at 600 rpm and applying a vibratory impulse of the same frequency to a number of metal strips that pressed against the head.

Tourette reported that the vibratory sensation was not unpleasant and that it induced lassitude and sleepiness. Regrettably, that appears to be about all it did. There is no report of any real, long-tem, benefit. The same, unfortunately applies to the Charcot shaker.

The placebo effect of such devices on impressionable patients can be very powerful, especially when they are prescribed by people of such authority as Charcot and Tourette. So far as paralysis agitans is concerned, it is not unlikely that patients suffering from continuous strong tremors would feel relieved if, for a time, they were so shaken about by an external force that they were unable to appreciate their own pathological vibrations. This is called symptomatic treatment and, while not rejecting it, scientific doctors recognize that it is no substitute for an attack on the basic process causing the disease. It seems that this particular placebo effect was not strong enough to justify continuing with the idea and the Charcot shaker is now no more than a dusty footnote in the remote medical literature.

As for the Tourette trembler, even an assiduous medical historian would be hard pressed to find any reference to it. The well-known Tourette syndrome is at last being elucidated. Functional imaging techniques using positron emission tomography (PET scanning) have suggested that it is due to genetically induced minor abnormalities in parts of the brain known as the basal ganglia. Family studies have confirmed the genetic basis. This basis for the condition is also supported by the fact that a similar pattern of symptoms occurs in the disease encephalitis lethargica (sleeping sickness) which affects this part of the brain in a specific fashion. In addition it is known that drugs such as haloperidol, which are effective in relieving the syndrome, act on the basal ganglia. The brain changes are subtle and, regrettably, there is no reason to suppose that they can be reversed by vibrating the brain. All the evidence we have on the effects of brain-shaking suggests that it is the last thing the brain needs.

> To imagine that pain can ever be abolished in surgery is a daydream that is absurd to try to realize. Pain and the knife are two words that every patient must forever live with. And we who are doctors must recognize the fact.
>
> *Professor Alfred Velpeau of*
> *the Paris Faculty of Medicine, 1839*

Rotten fish as the cause of leprosy

Sir Jonathan Hutchinson (1828–1913) was a surgeon at St Bartholomew's and the London Hospital, a professor of surgery, a President of the Royal College of Surgeons, and a Fellow of the Royal Society. He was a renowned authority on eye diseases and eye surgery, an expert skin specialist and a world authority on syphilis. Even today, every medical student knows about Hutchinson's teeth – a characteristic of congenital syphilis – but not all are aware that Hutchinson's name is attached to no fewer than 10 different conditions or symptoms. He was an indefatigable writer and produced numerous books, including the 10-volume *Archives of Surgery*.

But when Sir Jonathan delivered his retiring address on relinquishing the appointment of President of the College of Surgeons in 1903 more than a few eyebrows were raised. The theme of his address was his conviction that leprosy was caused by eating bad fish. For the previous two years, the great man had been visiting leper communities in Africa and India, and, as a true scientist anxious to determine whether there was any evidence against his hypothesis, had been seeking out especially those leper colonies where, as he had been told, fish was not consumed. In every single instance, Sir Jonathan found that his informants had been wrong and that fish was being eaten.

Hutchinson's theory of the causation of leprosy was well known at the time and his position as a medical expert was so strong that not a few people believed him. One scholarly friend had sent him a quotation from Erasmus (1466–1536) whose dialogue on rotten fish, *Ichthyophagia*, records that the Pope, himself, had proposed that preserved fish should be proscribed because of its tendency to spread leprosy.

There were a few snags about Sir Jonathan's theory. For a start, his contemporary Louis Pasteur (1822–95) had shown unequivocally in 1865 that infectious diseases were caused by 'germs' and that these spread form person to person. Moreover, in 1874, the Norwegian bacteriologist Gerhard Hansen (1841–1912) had isolated *Mycobacterium leprae*, the cause of leprosy. His careful microscopic examination of biopsy specimens taken from leprosy victims had revealed the typical rod-shaped bacteria, and his subsequent public health measures had reduced the number of known cases of leprosy in Norway from 1760 in 1875 to 575 in 1900. In 1897, Hansen was appointed chairman of the first International Conference on Leprosy and later was made president of the second conference.

Can it be possible that the great Sir Jonathan Hutchinson had stopped reading his medical journals?

> I contend that we violate the boundaries of a most noble profession when, in our capacity as medical men, we urge our fellow-creatures, for the sake of avoiding pain alone, to pass into a state of existence the secrets of which we know so little at present. What right have we to say to our brother man, 'Sacrifice thy manhood – let go thy hold upon that noble capacity of thought and reason with which thy God has endowed thee, and become a trembling coward before the mere presence of bodily pain'?
>
> *A Liverpool surgeon, attacking chloroform general anaesthesia for surgery, in 1847*

The syphilis scandal

This example concerns the research known as the 'Tuskegee Study of Untreated Syphilis in the Negro Male'. Tuskegee is a small city of about 13,000 inhabitants and the seat of Macon County in east central Alabama. It dates from 1833 and took its name from Taskigi, a nearby Indian village. Tuskegee is best known for its educational Institute, a co-educational college for black students. This was started as a primary and secondary school, but has since been elevated to university status and now offers degree courses in science, engineering, education, business, architecture, and other subjects.

There were certain indications of colour prejudice in the region. An attempt was made by the state legislature in 1957 to exclude most of the black voters from the electoral roll by changing the city boundaries. Happily, this was foiled by the US Supreme Court in 1960 who declared the move unconstitutional. The first black mayor, John Ford, was elected in 1972.

In 1932 the local Government health department made an extraordinary decision. This was that black men who were found to be suffering from syphilis would be deliberately denied any form of effective treatment. The idea was to chart the progress of the disease and to compare the health of the sufferers with black men who did not have syphilis. None of the men was aware that treatment was being withheld. Indeed, because they were attending a government medical facility, they all took it for granted that they *were* being treated. Antibiotics, now used with excellent effect to treat syphilis, were not generally available until about 1945, but, at the time the study started, syphilis was treated with organic arsenical drugs. Paul Ehrlich had won a Nobel Prize for the discovery of salvarsan (arsphenamine) in 1909, and an improved version, neosalvarsan (neoarsphenamine), was the standard treatment until the development of penicillin in 1943. Neoarsphenamine required a long course of injections – in some cases for two years – but produced full recovery and eradicated the infection.

Three hundred and ninety-nine black men were recruited for the trial, and, even when penicillin was plentiful, none received treatment. Even after serious doubts had been expressed as to the usefulness of the study, it went on. When it was finally stopped, as a result of protests in the 1970s, 100 of the men had died of syphilis, 40 wives had contracted the disease, and 19 children had been born with congenital syphilis.

The really horrifying aspect of the matter is that syphilis had previously attracted an enormous amount of medical interest and every detail of the disease was already thoroughly well known and documented. Because of its extensive medical effects and special pathological interest, it was, in fact, one of the best known of all diseases. There was, in the textbooks and journals, an enormous amount of clinical and pathological information about untreated syphilis and there is no reason to suppose that this study added anything to medical knowledge.

After ending the trial, the US government began to compensate participants with *ex gratia* out-of-court payments and free medical care. The government did not, however, admit that it had done anything wrong until 16 May 1997, when President Clinton, at a White House ceremony, formally apologized. Eight of the participating men were still alive and five of them attended the ceremony.

Psychology

> 'The heart is the seat of intelligence.
>
> *Aristotle (384–322 BC)*

The science of phrenology

The German physician Franz Joseph Gall (1758–1828) believed himself to be a true scientist. Writing to Baron Joseph de Retzer, he said: 'The purpose of my work is to find out the functions of the brain, especially its different parts, and, in particular, to show that it is possible by observing various elevations and depressions on the surface of the brain to determine the degrees of different aspects of the personality. This work will be of the first importance to medicine, morality, education and the law – indeed to the whole science of human nature.'

Gall's interest in the relationship of appearance to personality started when he was a schoolboy and happened to notice that all the fellow students who challenged his place in class had prominent eyes. This impressed him greatly and he soon became convinced that protruding eyes were a sure sign of talent. Further investigation of people with this feature persuaded him that the sign indicated verbal rather than general ability. Next he noticed that a particular set of the eyebrows seemed to be associated with a singular ability to find birds' nests and to remember where they were. He also observed that people remarkable for their determination had an enlargement in a particular part of the head.

These observations greatly excited Gall, who now determined to see whether differences in the shape of the head could be correlated to various human characteristics. So he begged his friend Dr Nord, who was the physician to a lunatic asylum in Vienna, to allow him to examine the heads of the inmates and to note the type of lunacy. He also frequented prisons and courts of justice and made careful notes of the shape of the heads of criminals. These observations were extended to anyone who showed a well-marked characteristic. He studied musicians, mathematicians, artists of various kinds, orators, greedy people, logical people, stupid people. Whenever such a person also showed an unusual shape of head, Gall made a careful note of the relationship.

Eventually, he concluded that people's personalities could be divided into 27 'faculties', which were innate and unchanging. The brain was the organ of the mind and each of these faculties was controlled by a particular part of the brain. If the faculty was well marked, its brain area would be well developed and would be prominent. The skull, he suggested, was so closely moulded to the brain that it formed an exact copy of the brain surface. Thus it was possible by an examination of the outer surface of the skull to form an accurate assessment of the whole character of the individual.

Gall's 'faculties' included such things as kindness, arrogance, sense of places, musicality, sexiness ('instinct of reproduction'), tendency to murder, sense of metaphysics, poetical talent, firmness of purpose, and memory for words. It is worth noting that Gall's 'instinct of reproduction', sometimes described as 'amativeness' was located in the region of the cerebellum – the hind brain – and that Gall's own personality was noted for its 'amativeness'. This point has a curious association, as will shortly be seen. Initially, Gall left blank areas between the various faculty bumps, but these were later filled in by others who followed his system. Eventually, there were 35 'faculties' and all the spaces had been filled in.

In 1796 Gall gave his first course of lectures on the new science that he had developed and, before long, he was famous. Some of his ideas were shocking, especially his claim that the brain was the organ of the mind. This aroused enormous controversy. Many protested that this idea was atheistic, fatalistic, materialistic and destructive of religious faith. The very bonds of society would be broken. Others enthusiastically accepted and took up his ideas, seeing in them the foundation of a new and valuable understanding of the human being.

Gall tried to avoid getting involved in these arguments and went on quietly collecting data. He and his pupil and supporter, Dr Spurzheim, continued to lecture and to pour out books on the subject of phrenology. It was Spurzheim who extended the 'organs of the mind' to a total of 35, namely: destructiveness, amativeness, philoprogenitiveness (love of offspring), adhesiveness (sticking to it), inhabitiveness (love of home), combativeness, secretiveness, acquisitiveness, constructiveness (creativity), cautiousness, approbativeness (prone to praising), self-esteem, benevolence, reverence, firmness, conscientiousness, hope, marvellousness (being easily impressed), ideality (given to thinking some things are perfect), mirthfulness, imitation, individuality, configuration (interest in pattern and arrangement), size, weight and resistance, colouring, locality, order, calculation, eventuality, time, tune, language, comparison and causality (concerned with cause and effect).

The two colleagues went on a lecture tour of Germany and

Switzerland and in 1807 they arrived in Paris. Here they met with greater enthusiasm – as well as stronger opposition – than they had encountered anywhere. Even Napoleon showed an interest at first, but later he decided that phrenology was nonsense and appointed a commission to look into it. Possibly influenced by Napoleon's known attitude, the commission produced a report highly unfavourable both to phrenology and to Gall. Undaunted, the two continued to lecture and write until 1813, when they quarrelled over the amount of credit each deserved for the current state of advancement of the science. Spurzheim went off to England to continue on his own and Gall remained in Paris until his death. A post-mortem examination showed that he had a small tumour of the cerebellum – in the area he had associated with 'amativeness'. Oddly, Gall's skull was found to be at least twice the normal thickness.

Phrenology had a remarkable vogue for many years and was widely believed to be an exact science by many, including scientists and doctors. People queued up to have their 'bumps' read and phrenology booths sprang up like mushrooms, usually organized by self-styled professors who had read one of Gall's books. Many people were convinced that they had not recognized their true characters and some, no doubt, tried to live up to the qualities their phrenological analysis suggested they ought to have. This could be good news or bad news.

Phrenology was even bigger in America than in Europe. Two of the most enterprising New World exponents of the science were the Fowler brothers, Orson and Lorenzo, who kept phrenology flourishing in America long after it had passed away in Europe. In 1838 they even founded and ran the *American Phrenological Journal*, which actually remained in publication until 1911 by sensitively detecting and responding to the wishes of the American public. The Fowlers were ready to read anyone's head – for a fee – and ran a thriving and profitable business as advisers on anything from the choice of a mate to the most senior business appointments.

We now know that phrenology is complete nonsense. For a start, although the brain does fit tightly inside the skull, it is not the firm, almost hard, organ that a formalin-preserved specimen in an anatomical museum would suggest. The living brain is soft and jelly-like. It is quite true that different parts of the brain have different functions – this is called cerebral localization – but these are functions such as movement, sensation, vision, hearing, sense of smell, speech and so on. There is no single part of the brain responsible for memory or for the qualities that Gall and his followers claimed.

The 'science' of phrenology did, though, have a positive side. In 1861 the French surgeon Pierre Paul Broca (1824–80) examined, *post mortem*, the brain of a woman who had lost the power of speech some years before after a stroke. He found an area of destroyed tissue in the left side of the brain. Broca concluded that this was the brain area for speech and that there were probably other areas serving specific functions. Broca's idea was violently rejected by his contemporaries, but the observation led others to see whether, in other cases of brain disease and injury, there was evidence of localization of function. Soon such evidence was found, and, by the end of the nineteenth century, many of the functional areas of the brain were known. This was the beginning of the modern science of neurology.

It is to Gall's credit that he not only insisted that the brain was the seat of the mind, but that he was also the first to suggest the idea of localization of function and, perhaps, made the idea more acceptable. This is often the way that science advances – by proposing ideas that, while wrong in detail, contain an important element of truth.

> Until there is a practical alternative to blind trust in the doctor, the truth about the doctor is so terrible that we dare not face it.
>
> *George Bernard Shaw (1856–1950)*

The Anna O. blunder

In his paper 'Project for a scientific psychology', Sigmund Freud said: 'The intention is to furnish a psychology that shall be a natural science; to represent psychical processes as quantitatively determinate states of specifiable material particles.' Freud thought that he could achieve this ambitious aim. Right to the end of his life he insisted that psychoanalysis had the status of a natural science and that his work had placed psychology alongside other disciplines, such as physics and chemistry, with equal scientific status. Here we examine whether this is true.

The catch-phrase and rallying cry of modern scientific medicine is that it is 'evidence-based'. In other words, a method of treatment should be considered valid and should be adopted only if there is good evidence that it works. This is not quite so simple as it may seem. If a doctor gives a patient a particular treatment and the patient recovers from the illness, the doctor is not entitled to say that this is evidence that the treatment brought about the recovery.

Fortunately for all of us, we have highly efficient, built-in, treatment facilities that automatically set about correcting disorders and healing disease. These include mechanisms to heal wounds and a remarkable infection-combating resource called the immune system. So treatment of organic disease is, in most cases, unnecessary and can be left to the body's own processes. Any treatment that might have been given may possibly have helped, or it may have had no effect at all, or it may even have tended to make things worse. In all three cases, recovery is likely and if this does occur, the chances are that the treatment had nothing to do with it. Looked at in this way, the doctor's claim to have cured the patient is as best dubious and at worst ridiculous. This kind of logical process is, however, commonplace and doctors and other therapists are often only too pleased, by implication, to take the credit. Incidentally, the 'it must have been the treatment' kind of reasoning is what keeps alternative medicine alive and well. The Romans called it the *post hoc, ergo propter hoc* fallacy.

So how does evidence-based medicine work? Well, for a start, you need a large number of cases – at least several hundred. Half of these have to be given the treatment and half – known as controls – get a dummy treatment that cannot possibly do any good. The controls have to be properly matched to the people being treated. Then you have to make sure that the patient does not know whether he or she has had the treatment or is a control. This is because knowing you have had a treatment is enough in itself to make you feel better, or even in some cases, to get better (the 'placebo effect'). It is also necessary for the doctor not to know which patients have had the real treatment and which the dummy. This is because, however honest he or she may be, it is hard to avoid unconscious differences of behaviour towards the two classes of patients. Only at the end of the trial, when all the results are known, is the truth revealed. Since both the patient and the doctor are unaware of what is going on, this kind of trial is called a 'double-blind' trial.

How does Freud's claim to be a scientist match up to these criteria? And on what sort of basis did Freud found the whole practice of psychoanalysis? How many patients were involved? Was the placebo effect of the treatment eliminated? Did he, the doctor, provide any controls?

Bertha Pappenheim (Anna O.), a highly intelligent, well-educated, strong-willed, poetical and imaginative, but moody, young woman was 21. Her family was puritanical and she led a rather boring life looking after the house which she enlivened by conducting an active imaginary life – what she called her 'private theatre'. In July 1880, Anna's father developed an abscess in his pleural cavity and was very ill for several months. Anna was passionately fond of her father and was deeply distressed by his illness. She nursed him devotedly, so much so that her own health was affected. She became weak, lost her appetite, developed a cough and was found to be anaemic. To her distress, she was no longer able to nurse her father. Each afternoon she found it imperative to rest and in the evenings she passed into a hypnotic-like state.

At the beginning of December Bertha developed a marked in-turn in one eye – a 'convergent squint' – which caused her double vision, and on 11 December she took to her bed. She remained bedridden until April of the following year. During that time some alarming symptoms developed. She had a severe headache on the left side at the back; she felt that the walls of her room were tilting over; and she developed a paralysis of the muscles on the front of her neck, so that she could not move her head except with her hands. Then there was total loss of sensation and tightening of the muscles in her right arm and then in her right leg. The same then happened to her left arm and leg. Her legs were extended, pressed together and her feet turned inwards; her shoulder joints became completely rigid; and, perhaps unsurprisingly, she became remarkably anxious. Later she developed a severe disorder of speech in which her command of language deteriorated progressively until she became almost unintelligible. Sometimes she would use four or five languages simultaneously. Eventually, for a period of about two weeks, she was unable to speak at all.

In addition to the physical effects, Bertha's personality showed some remarkable changes. It would alternate between a fairly normal state of depression and anxiety and a condition in which she was aggressive, abusive, violent, destructive, hallucinated – seeing black snakes instead of her hair – and complaining of becoming blind and deaf.

Bertha's doctor was Dr Joseph Breuer, a great friend of Freud's. Breuer quickly decided that these symptoms could not be the result of organic disease. He inferred that she had become 'very much offended' over something about which she could not speak. Breuer then exerted strong persuasion and eventually Bertha came clean. As she did so, the 'inhibition' that had made it impossible for her to speak disappeared. At the same time, the paralysis on her left side recovered. From that point, this Viennese girl spoke only in English, apparently without realizing it. It was some months before Breuer could persuade her that she was not talking in German. She perfectly understood people who spoke to her in German, however. During the time she was confined to bed she rarely saw her father and then only for very short periods.

A notable symptom was an apparent deafness. She would be unable to hear someone coming into the room. If her body was shaken she would become deaf. If she was frightened by a noise or if she listened hard for a long time, she would become deaf. On 1 April 1881 Bertha was able to get out of bed, apparently completely recovered from the paralysis on the left side, but still weak and insensitive on her right side. Five days later her father died. For two days Bertha was in a kind of stupor, from which she recovered much changed. She was very quiet and seemed much less anxious. Her field of vision was now so reduced that when she looked at a bunch of flowers she could see only one blossom at a time. She said she could not recognize people. To identify people previously well known to her she had to analyse one feature at a time. People, she said, were like dummies in a museum and seemed to have no connection with her. The only person she could recognize immediately was Dr Breuer.

Bertha continued to speak only in English and now was unable to understand anything said to her in German. She could read French and Italian but, when reading aloud, translated what she was reading fluently into English. Having been right-handed, she now wrote only with her left hand using capital letters. At this point Bertha refused to eat anything herself, but would allow Breuer to feed her. Once Breuer was absent for a few days and when he was away she refused all food and had violent hallucinations of skeletons and other terrifying figures.

Breuer had discovered that if he visited Bertha in the evenings when she was in her 'hypnosis state' and got her to talk about what she had experienced during her bad afternoons, she was greatly relieved. In fact, talking in this way made her feel so much better that she described the procedure as her 'talking cure' – significant words. A year after her father had died, Bertha was living in a different house but was still having regular alternating states. These were now occurring more often during the day. Now, when in her second personality, she believed she was living exactly one year before and was back in the year in which her father had died. During these phases she had no memory of contemporary events. When in the 'previous year' she behaved as if she were in the room she had occupied then, even to the extent of walking into a stove that occupied the position of the door in her old room. She would tell Breuer she was angry with him and he was able to confirm from his diary that he had annoyed her on the same evening the year before.

Breuer then discovered that the period between the onset of her father's illness in July and the time at which she had taken to her bed, in December, was critical. When she was encouraged to talk about all the events of that period, her symptoms disappeared. Even more significantly, each symptom appeared to be connected with a particular event and was permanently abolished by her talking about that event. When he realized this, Breuer took each individual symptom and, working backwards, during the critical period, eventually unearthed the event that had originally produced it. To facilitate this, Breuer would hypnotize Bertha, which he achieved very easily.

In this way he was able to show that Bertha's inability to hear when someone came in could be traced to her having failed to hear her father come in. Her deafness, which was repeatedly brought on by her being physically shaken as in a carriage, was traced to her having been shaken by her brother when he found her listening at her father's sickroom door. Breuer, now fascinated with the case, proceeded systematically to try to remove all Bertha's symptoms. At this point, Breuer's wife informed her husband that he was spending far too much time with the young woman, and Breuer told Bertha that he had to go off on a trip with his wife. That evening he was called back urgently to see Bertha and found that she was claiming to be in love with him and to be having his baby. This was, happily, a phantom pregnancy but the girl's transference (sexual infatuation) on to Breuer was so powerful that the doctor became thoroughly alarmed and gave up his investigative activities.

Breuer wrote up the case in detail and claimed to have effected a complete cure. He did not publish this case history for many years, but he showed it to Freud, thereby originating the practice of psychoanalysis which Freud took up enthusiastically and developed enormously. Incidentally, Bertha was not cured by Breuer's 'talking cure' and she had to have treatment subsequently in the Bellevue Sanatorium in the Swiss town of Kreutzlingen, still losing her ability to talk German and becoming addicted to morphine. The case notes from the sanatorium were found many years later and these showed that Bertha was suffering not from hysteria but from the brain infection tuberculous meningitis. Many, if not all, of her symptoms could be adequately accounted for by this serious condition. Her father's pleural abscess was almost certainly tubercular and Bertha must have been regularly exposed to infection as she nursed her father.

So much for the effectiveness of the 'talking cure' on which psychoanalytic practice was based. What of the number of cases on which the validity of the new treatment was based? Breuer and Freud collaborated on the book *Studies on Hysteria* in which the 'talking cure' was described as a definitive form of treatment. In addition to Breuer's case of Anna O. (Bertha), this book deals with four other cases, with no controls. With each of these cases, Freud developed and elaborated the procedure of psychoanalysis further.

Katherina was a girl of 18 who approached Freud when he was on holiday high in the Alps. She told him that she suffered from suffocating breathlessness of sudden onset accompanied by a crushing feeling in her chest. Her throat was tight and she felt as if she was going to choke and die. Freud asked her if she was thinking of anything when the attacks came on, and she said that she always saw a frightening face looking at her. Freud's questions elicited that she did not recognize the face and that she had been having the attacks for about two years. He asked her whether anything embarrassing had happened to her just before the attacks started. 'Oh yes!' she said, 'It was when I caught my uncle with Franziska, my cousin.'

The story was that Franziska, who was required to do the cooking, could not be found. The aunt was away and Katherina and her brother searched the house and decided that Franziska must be in their uncle's room. So they went there and found the door locked. Katherina then peeped through a window into the darkened room and saw that her uncle was lying on top of her cousin. At that she suddenly developed an attack of suffocation. Freud asked her about the face she saw and suggested it might be that of her uncle. The girl naively enquired why her uncle should be making such a dreadful face just then.

Katherina then related how eventually she told her aunt what she had seen. The result was a succession of dreadful rows in which Katherina learned a great deal about sex. Meanwhile Franziska was found to be pregnant and the aunt left her husband. Freud continued to question Katherina and to his surprise she told him that her uncle had made several sexual advances to her, starting when she was only 14. According to Freud's report in the book, Katherina's detailed account of these events transformed her: from being sulky and depressed, she became cheerful, bright and exalted.

These five cases were the 'scientific' basis for the practice of psychoanalysis. Although Freud assembled a great body of theory to back up psychoanalysis, he gave no scientific evidence to support his assertions. Now no one will deny that Freud was one of the most influential figures of the twentieth century. The effect of his writings on human opinion, attitudes and even behaviour is incalculable. But his claim that psychoanalysis is a science must be questioned. Psychoanalysis is a pseudoscience, many of the assertions of which cannot be disproved. It also contains testable theories which, when disproved, are still upheld. As Frank Cioffi points out: psychoanalytic theory contains numerous peculiarities which are apparently unrelated but which are, in fact, manifestations of the need to avoid refutation. 'It is characteristic of a pseudoscience,' writes Cioffi, 'that the hypotheses which comprise it stand in asymmetrical relation to the expectations they generate.' They are, he explains, allowed to be vindicated if they are fulfilled, but not to be discredited if they are not.

By any standards Freud was a remarkable man. He was a brilliant scholar with a thorough working knowledge of Latin, Greek, English, French, Italian and Spanish and of the literature of these countries. As a schoolboy he had been steeped in the classics and, as his writings confirm, his knowledge of mythology was profound. He was also a man of great courage. In our socially enlightened days it is hard to appreciate the strength of the public outrage expressed against anyone who, in Freud's time, had the temerity to talk openly about sex and its manifestations. It is also hard to gauge the moral courage required of a man of serious purpose, in these repressive times, to face up to the accusations, implicit or expressed, of dirty-mindedness.

But Freud was equal to these challenges. Joseph Breur was not, for when he saw the necessity to acknowledge the central role of sexuality in human motivation, he turned away. Freud alone was left to delve into this murky region and to demonstrate unequivocally and openly the importance of sex in the human condition. It was that contribution that made Freud so enormously influential on twentieth-century thought. Freud's blunder was to think that he was a scientist. He was not. He was a literary man of the greatest importance.

There is no scientific evidence to support Freud's claims of the Oedipus complex, or the universal death wish, or his insistence that the mind was a tripartite entity consisting of the id, the ego and the superego. His dream symbolism, in which every elongated object represents a penis and every container a vagina, was no more than an effect of a repressive age in which the frank acknowledgement of such anatomical features was unthinkable. The hysterical manifestations of repressed sexuality that seemed to be so common in his time hardly exist nowadays. Freud's ideas were a product of his times: they are not eternal truths, but merely a feature of a particular cultural trend and a particular stage in social evolution.

One unhappy consequence of the non-scientific nature of psycho-analytic theory was that it quickly turned into a religion. It developed a set of dogmas that it was heretical to question. It developed a hierarchical priesthood. It became rent by schisms and distressed by heresies. Jung broke away and founded his own religion. Adler broke away and set up his own sect. Wilhelm Reich and Melanie Klein broke away. Passions ran high. Freud's desire to preserve the purity of his original conceptions meant that he made an enemy of anyone who questioned them.

Like all religions, psychoanalytic theory became fragmented. There are now scores of 'schools' of psychotherapy. By definition, only one of these can be scientific, and it is questionable whether that one has yet emerged.

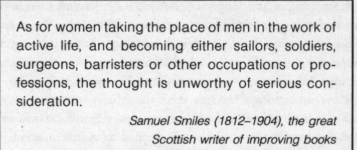

As for women taking the place of men in the work of active life, and becoming either sailors, soldiers, surgeons, barristers or other occupations or professions, the thought is unworthy of serious consideration.

Samuel Smiles (1812–1904), the great
Scottish writer of improving books
and promoter of self-reliance and hard work

A great man loses touch with science

The Swiss psychologist Carl Gustav Jung (1875–1961) is regarded by many as one of the most notable men of the twentieth century. At the time of his death at the age of 86, he was widely believed to be the greatest living psychologist. He was the founder of analytic psychology, the originator of the idea of extroverted and introverted personality, of archetypes, and of the collective unconscious. Like his one-time colleague and associate Sigmund Freud, Jung practised a profession based explicitly or implicitly on science but, in fact, his interests and activities extended far beyond science. He is more properly described as a free-ranging thinker, a promoter of philosophical concepts, almost an artist in ideas.

Reference to a man of Jung's status in a book of this kind can deal only narrowly with one aspect of the man. The question to be addressed is how far he can be considered to have been a scientist. This is important because it has a bearing on how far we should be influenced by, or even accept, his conclusions. It is also important because of the light it casts on the nature of science.

From early in life, Jung was intensely religious and there was an assumption that he would become a pastor, following in the tradition of his father and other members of the family. But as a teenager he discovered philosophy and read widely, and this led him to abandon the idea. His religious faith remained with him throughout his life, however, and strongly influenced his thought. At the age of 84, less than two years before he died, he was asked whether he believed in God. 'I know,' he answered, 'I don't need to believe. I know.'

Jung spent his life in search of the soul – a concept about which science has nothing to say. His break with Freud came, ostensibly, over Freud's insistence that sexuality was the principal factor in the causation of neurotic disorders. Jung was well aware of the importance of sexuality, but he saw libido as something much wider than just sex. To him it was the totality of human energy, the life force itself. But Jung's separation from Freud had a more fundamental basis than that. Freud genuinely believed himself to be a scientist. Whether Jung did or not is unclear, but he was engaged in a scientific discipline and repeatedly made claims to have discovered 'truths'.

Freud's ideas, however outlandish, at least showed some respect for the laws of causality. The same cannot be said of those of Jung. The idea of the collective unconscious – the part of the unconscious mind derived from ancestral memory and shared by everyone – was developed to explain a number of anthropological and other observations. But this was an idea that transcended science. It was, in fact, mystical, and fitted more easily into the category of theology than into science. Much the same goes for his theory of archetypes, primitive mental images inherited from our ancestors and present in the collective unconscious. These, he believed, were basic to the study of the psychology of religion. Jung's archetypes were universal patterns that he saw everywhere, expressed both in images and in behaviour.

He was a scholarly man who was constantly coming across links and associations in literature that seemed to him of the highest significance. He had a taste for obscure and often neglected texts of ancient writers, in which he would find matter that illuminated both his own dreams and fantasies and those of his patients. He even went so far as to insist that psychotherapists could never become truly successful practitioners unless they were familiar with writings of the old masters.

Late in life Jung began to suggest that while the laws of nature continued to apply to the physical world, they did not apply to the world of the psyche. He postulated a non-causal principle which, he suggested, allowed for psychic phenomena, prophetic dreams, clairvoyance and telepathy. He was deeply impressed by his observation that symbols used by the alchemists could often be found in modern dreams and fantasies. He believed that the alchemists had, in effect, put together a kind of manual of the collective unconscious, and he devoted four volumes to proving it.

Religious belief can, of course, coexist with strict scientific principles. But if it does and an attempt is made to achieve a synthesis of the two, critics are entitled to point out that the activity can no longer be described as scientific. Jung's critics have done exactly that. They suggest that Jung broke faith with the scientific method and returned to an earlier stage of superstition involving belief in magic, alchemy and astrology. This, some suggest, is the stage from which only science has been able, by slow and painful labour, to drag humankind into the light.

Jung would not have seen it that way.

> In future, the largest touring cars, if of reciprocating type as to engine, will not be of more than 1000 cc.
>
> *Professor A. M. Low*

Hans Selye and the effects of stress

The idea of stress, and its probable effects on health, has long since caught the public imagination. Books, magazine and newspaper articles, and TV programmes continue to pour out the message: 'Stress is bad.' But is this true? Stress causes secretion of adrenaline. The heart beats more quickly, the blood pressure rises, breathing becomes more rapid, the muscles are tense, and there is a feeling of 'butterflies in the stomach'. As we shall see below, these physiological responses evolved to promote survival.

The idea of stress in its modern sense was introduced in the early 1950s by the Austrian-born Canadian physician Hans Selye (1907–82). Selye's English wasn't very good and, as he later admitted, he was a bit confused about the difference between stress and strain. Up to that point we had all been familiar with the word 'strain'. 'Poor John,' we would say, 'He's been under terrible strain.' Selye pulled off a remarkable double. By using the word 'stress' when he meant what everyone else meant when they said 'strain', he gave the lay public the idea that he was talking about something new. The term 'stress' had been unfamiliar to anyone but engineers or physical scientists, but suddenly everyone was talking about 'stress'. The other thing Selye did was to use what was actually the correct term. The terms 'stress' and 'strain' come from engineering and are still often confused. Stress is the force exerted on a body that tends to cause it to deform, while strain is a measure of the extent to which a body is deformed when it is subjected to stress. A strain gauge is a simple device that measures the amount of the deformity.

The terms can, of course, be applied to human bodies in exactly this mechanical way. A bone can be stressed and can bend a little, thus exhibiting strain. But when we talk, these days, about a human being experiencing stress, we are not thinking of the application of a physical force that causes a bit of body squashing (strain). Our usage is purely metaphorical.

Selye studied medicine in Prague, Paris and Rome before moving to McGill University, Montreal, Canada. There he set up an Institute for Experimental Medicine and Surgery, and in 1945 he became its director. Selye was a physiologist familiar with all the existing knowledge about the hormonal changes that occurred in the body under conditions of anxiety. It had long been known that the production of adrenaline and the steroid hormone cortisol were necessary for survival in situations of 'fight or flight' in response to fright. During the course of human evolution, when life was 'nasty, brutish and short', only those young individuals capable of mobilizing these aids to alertness and sudden physical exertion would survive to breed. So, by natural selection, the production of these hormones became part of our physical and physiological make-up.

Selye first thought of the idea of biological stress when he was a medical student. While considering the causation of illness, it had occurred to him that there was one thing that all ill patients had in common. However diverse their diseases, they all looked and felt sick. Was it possible that some general process was operating that had this general and universal effect? Excited by this idea he went to see the professor of medicine and was promptly told to read his textbooks and forget such naive and childish nonsense. Of course they all felt and looked sick: they *were* sick. But Selye knew that some of the great ideas in science sounded, on first hearing, like childish nonsense, so he tucked the idea away for future reference.

Ten years later, while working at McGill, Selye made a discovery that excited him greatly. He found that rats who were exposed to various unpleasantnesses – given injections of mildly toxic substances, or kept caged out in the cold of winter, or persistently overworked – all showed enlargement of the hormone glands that produce adrenaline and cortisol. These are the adrenal glands, then known as the supra-renal glands because they sit on top of the kidneys. Even more interesting was the fact that many of these rats developed stomach ulcers. They were showing a general, and identical, reaction to a variety of stress-producing events. This tied up nicely with his earlier idea. Could this be the thing that all ill people have in common?

It seemed obvious to Selye that stress, whatever its nature – physical threat, actual injury, bacterial infection, social or marital problems, perceived danger of any kind, and so on – could cause various general effects. The 'fight or flight' stage he called the alarm stage. But this, he thought, was followed by an adaptation stage in which the body gradually got accustomed to alarms and developed some measure of resistance to them. If, however, the stresses continued, the person would pass into the third stage of the adaptation syndrome, the stage of exhaustion. This was the stage at which people developed ulcers, heart disease, rheumatoid arthritis, colitis, diabetes, or any other disorder to which they might be predisposed by heredity.

Selye published his findings and ideas in the medical press in the form of a theoretical proposal. He was, in fact, claiming that there existed a physiological/pathological process – the general adaptation syndrome – that had never previously been recognized. The doctors showed little interest, so Selye decided to write a popular book about it. This book, *The Story of the Adaptation Syndrome*, came out in 1952 and met with a strong public expression of approval. People found that Selye's ideas struck a resonant chord. They made sense and they seemed to fit in with personal experience. So further books appeared: *The Stress of Life* (1956), *From Dream to Discovery* (1965), *The Case for Supramolecular Biology* (1967) and *Stress without Distress* (1974).

Selye's account of the 'stressors' that caused all the trouble also struck a chord with the public. These included frustration, anxiety, conflict, alarm – all the things we think of as the 'stresses of modern life'. People were personally familiar with these and could rate them by the strength of the physiological effects they produce – often by how much muscle tension was felt in the upper part of the abdomen. They also included physical assault on the body by bacteria, injury, radiation, poisons, drugs, overcrowding, pollution, and so on. One thing perceived as being a potent stressor was frustration. Humans are goal-seeking animals and hate to be thwarted in their motivation. Frustration causes a strong emotional reaction and, since motivation covers the whole spectrum of our desires, none of us is free from frustration. This may be a succession of pinpricks or a life-long process, especially in the case of people whose goals are unrealistically high in relation to their innate abilities.

Selye's books soon made him and his ideas famous, and spawned a whole new popular literature. But they were distinctly unpopular with the medical profession, as the ideas in them had been brought into the public domain before they had been exposed to the criticisms of the profession – a process known as 'peer review'. Unfortunately, the sheer volume of popular books on this subject has given it validity in the eyes of the public.

The doctor's point of view was that stress was an integral part of life, a normal feature without which we could not survive. In acute emergencies, people could perform amazing feats of strength or agility that would be impossible without stress. They believed that there was misunderstanding of what was meant by stress. It was not, for instance, necessarily unpleasant. Many people enjoyed it, thrived on it and even sought it. Playing, or even watching, competitive games, involved considerable stress. In some circumstances, stress was necessary to achieve a certain minimal standard of performance. Actors, concert pianists, and surgeons, needed a bit of stress to give of their best. There was little or no evidence that the ordinary stresses of living or even long-sustained unpleasant stress need have any effect on health. The idea that it would inevitably cause heart attacks, strokes, ulcers, skin disease or cancer simply was not borne out by the facts.

There was so much hype about stress that everyone came to think that there must be something in it. The word became ubiquitous and even began to appear in the medical literature. 'Soft' science – the kind that relies on the sort of assertive rather than proved 'evidence' that Selye was putting forward – had a field-day with stress. Two psychologists, Thomas Holmes and Richard Rahe, of the University of Washington, produced tables of stress from everyday life that rated events in terms of the severity of the stress they might be expected to produce. They arbitrarily allotted the figure of 100 to a very strongly stressful life event, such as the death of a spouse, with appropriately smaller scores for lesser stressors. Categories covered included such things as being sacked from a job, moving house, having a row with the boss, getting married, getting divorced, having a baby, loss of various relatives, and so on. This is the kind of thing that seems, on the face of it, interesting and valid, and many people accept such suggestions unquestioningly. However, such tables imply a quite unrealistic degree of uniformity in response from human beings. Obviously, for different people, different events will have widely different importance and different stress values. To take an extreme example, while the death of a spouse will, in the great majority of cases, cause severe distress, there will always be cases in which it will cause rejoicing. As we move down and away from such an extreme case, the lack of uniformity in response to the stressors becomes greater. A row with the boss, for instance, will generally be thought stressful but could range from a minor disagreement, that ends with the boss respecting you more, to an encounter that ends in your losing your job. Moving house may involve, in terms of stress, anything from deep satisfaction to profound anxiety. People are far too varied to be usefully categorized in their responses to such events.

The stress industry, inspired by Selye's books, gave rise to another piece of soft science – the theory of A-type and B-type personalities. This one appeared in 1974, the brainchild of two heart specialists, Meyer Friedman and Ray H. Rosenman, and described in the book *Type A Behaviour and Your Heart*. These cardiologists, who were especially interested in the causes of heart disease, suggested that many people create their own stress. Those who do are the A-type people – the fast-lane, highly competitive, impatient and hard-driving workers who do everything in a great hurry. A-type people are the ones who have a harrying sense of the urgency to time everything; they cannot bear traffic jams; they are engaged in a persistent struggle with themselves and demand perfection of themselves in everything they undertake; they are always early for appointments; they take on more work than they can comfortably manage; and they display a well-rationalized hostility. B-type people are calm, laid-back, easy-going, relaxed and patient.

The two heart men concluded that type-A behaviour was a better predictor of heart attacks than almost any other combination of factors. B-type people were much less prone to heart attacks. This concept seemed so plausible that nearly everyone outside medicine, and quite a few people inside the profession, immediately accepted it. For a time, it even featured occasionally in the medical literature. So a remarkable situation evolved. There was great enthusiasm outside the scientific world for the stress theory of disease, and equally enthusiastic scepticism and rejection inside it. Selye's ideas were strongly supported by uncritical journalists and others who accepted the whole thesis, and largely condemned by doctors and physiologists. Selye died without ever having gained general medical acceptance for his ideas.

Why were the scientists and the doctors so obdurate? There were several reasons. For a start, it is not acceptable to base a theory on a concept (stress) identified only by name but not clearly and unequivocally defined. Stress does this, stress does that, but what actually is stress? Nearly all the writing on the subject seems to assume that stress is some kind of noxious agent, external to humans, that operates harmfully upon them. That it is, in effect, analogous to a virus or a blunt instrument. Whatever it is, it certainly is not that. Stressors are not, in themselves, stressful. Stress is a feeling, an entirely subjective experience. It is an emotion experienced as a result of an interaction between a person and the environment, using this word in the widest sense. The result of such an interaction varies enormously from one person to another. One person's stress is another person's exciting challenge. Stress is in the eye of the beholder. The stress literature is notable for its confusion between cause and effect.

Once this point is made it becomes difficult to sustain the simplistic proposition that stress is a significant risk factor for disease.

When we consider the wide individual differences in response to stressors, we see that the tendency to develop disease can be just as validly attributed to individual predisposition from heredity or plain bad luck. We are all exposed to stressors, we all experience the stress emotion, and we all cope with it in different ways. Some cope better than others. 'The fault, dear Brutus, lies not in our stars but in ourselves.'

Critics of Holmes and Rahe's life-change stress factor theories pointed out that people predisposed to physical or psychological disease may be just the kind of people who are liable to get involved in 'stressful' changes. People liable to illness often work less well, are liable to dismissal, have to change homes, and so on. Long-term partners and spouses share environmental influences that commonly lead to the development of similar disorders. Such people may come from families with a higher than average incidence of bereavement. And so on. These criticisms seriously reduced the scientific status of this theory.

It is now believed that it is far too simplistic to divide people into A-type and B-type personalities. As with another proposed categorization, Jung's division of humankind into introverts and extraverts, there is a spectrum of personalities. Of course there are people at both extremes of the spectrum. A small percentage are obvious A-types or obvious B-types, just as there are obvious introverts and extraverts, but the great majority of people are neither. This makes the entire concept rather pointless. Nearly all the 'evidence' for linking A-type personality with heart disease is in the popular literature.

The term 'stress' seems, regrettably, to be cropping up far more frequently in textbooks and medical papers than ever before. This does not imply that the doctors now believe in the stress theory of disease. But it does imply that doctors have been affected by the massive force of public opinion on the matter. There are strong pressures on doctors to acknowledge that stress may be a significant contributor to disease. This is not a healthy development. Pressures of this kind are apt to lead to a loss of scientific detachment and objectivity – factors essential for the disinterested pursuit of science. It is a substantial scientific blunder to pay any attention whatsoever to lay pressures of this kind. A single incontrovertible fact carries enormous weight; a mountain of dubious speculation, unsupported hypotheses, and journalistic hype must be allowed to carry no weight whatsoever.

Starting from an incontrovertible fact, we can make some progress in considering the general stress scene. It is an incontrovertible fact that there are levels of stress so severe that many people exposed to them suffer psychological damage. The post-traumatic stress syndrome is a real entity. Note that here we are considering extreme degrees. Extremities of all kinds tend to be dangerous. A little of what you fancy may do you good, but an extremely large dose may be disastrous. Even in an extreme context, individual variations remain an essential factor. If a trainful of people hits another train with heavy loss of life, all suffer psychologically but only a small percentage develop the post-traumatic stress syndrome with its repetitive reliving of the stressful event or events, intrusive and vivid flashback memories and severe nightmares. Only a few develop loss of memory of the terrible event. Only a very few suffer permanently from the disorder.

Moving backwards from such extreme cases, individuals who respond badly to stress may show certain symptoms and signs indicating that they are being exposed to levels of stress that could be, for them, dangerous. Such people may need help and counselling. They are people who, conscious of long-term stress, are becoming increasing irritable, prone to sudden anger, loss of concentration and increasing difficulty in making decisions, who are sleeping badly and are definitely off their food, who cannot relax and who react dramatically over trivial matters. Even more significant are those who become totally withdrawn and despairing, or who show profound depression.

One of the most horrifying scientific blunders of all time was the total failure, during World War I, to recognize that there were stress levels beyond which some people no longer had any defence. In that terrible war soldiers were exposed to appalling stress from long periods of intense artillery and mortar bombardment and small arms fire. These unfortunate men were frequently required to face almost inevitable death. They were ordered to get up out of their trenches and run across open land in the face of intense machine-gun and artillery fire. Those who broke down were said to suffer from 'lack of moral fibre'. Those who ran away were tried for cowardice and shot. It may be suggested that this was a military necessity and that the generals were well aware that some would be constitutionally unable to take it. But most of the doctors subscribed, at least tacitly, to the notion of 'lack of moral fibre'. Millions faced death without protest and were killed. Of those who survived, many thousands suffered for the rest of their lives from what was then called 'shell-shock' and what we now call the post-traumatic stress syndrome.

Fortunately for all of us, situations of the extremity that these men had to face occur very rarely and, by the application of good science and rational thought to the problems of humanity, and critical scepticism towards the forces of superstition, unreason and prejudice, will, we hope, become rarer still.

The Rorschach blot

Hermann Rorschach (1884–1922) was a Swiss psychiatrist and neurologist who invented a psychological test that quickly became famous all over the world. The Rorschach ink-blot test owes a good deal of its popularity to its game-like quality and its resonances of childhood, when most of us experimented with ink or paint blots.

Psychologists use a standard set of 10 cards each having a bilaterally symmetrical blot of the kind produced when a piece of paper with ink on it is folded in half. Some of the standard Rorschach cards are coloured, some are black, white and grey. The test in which the cards are used is known as a projective test because, when the subject is asked to say what he or she sees, the answer is said to be a 'projection' of that person's personality.

After it was introduced in 1921, the Rorschach test quickly became popular with psychologists and psychiatrists and rapidly became, with many, an almost standard clinical tool. Since then, millions of people have been tested using the Rorschach cards and, on the basis of their responses, millions of assessments of personality, intelligence, cultural backgrounds, level of anxiety, grasp of reality and even psychiatric diagnoses have been made. Especially popular has been the diagnosis of homosexuality, latent or overt, which is almost invariably made when the subject's response indicated that he or she is interpreting ambiguous features in the blot as sexual organs or characteristics of the same sex.

Any competent psychologist or psychiatrist should, after an hour's conversation, be able to make a valid assessment of a person's intelligence, cultural background, level of anxiety or grasp of reality without resorting to the Rorschach test. Careful research into the validity of the test has shown that, in many, if not most, cases, the assessments made are naive and simplistic and that many mental health professionals fall into the trap of making standardized judgements on the basis of similar responses to the blots.

In particular, research has shown that the responses that have been taken to be indicative of homosexuality are not, in fact, made more often by homosexual than by heterosexual people. They are instead the kind of responses that heterosexual people would naively expect homosexual people to make: they are a projection of the examiner's simple-mindedness rather than of the subject's sexual orientation.

Although research has demonstrated the gross unreliability of the test as a valid clinical tool in the hands of most therapists, the test is still in use. In vain, responsible textbooks of psychology warn against the dangers of reliance on this singularly useless and misleading procedure. It is estimated that about six million Rorschach tests are still given every year. It remains especially popular in the USA.

> After I invented the ophthalmoscope a distinguished surgical colleage told me that he would never use it as it would be too dangerous to admit light into the diseased eye.
>
> *Hermann von Helmholtz (1821–94)*

Memories are made of this

Memory is probably the most important attribute of a human being. Indeed, it is so important that when it is totally lost, as in advanced dementia such as severe Alzheimer's disease, the affected person is apt to be described as a 'vegetable'. A well-stocked memory of reliable and readily accessible information is really what we mean when we talk of an 'educated person'.

For many purposes it is most important that we can rely on the accuracy of memory. This is especially so in the case of professional activities such as the practice of medicine, the law, science and academia. In such contexts people pride themselves on the accuracy of their memories; if they are shown to be liable to inaccuracy, their professional status can be severely damaged. Fortunately for such people, factual information is usually built into a large matrix of knowledge that has the effect of making accurate recollections much easier. A scientist, for instance, will easily remember facts correctly if they are such that inaccuracy would immediately sound a discord with the general scheme of knowledge.

There are, however, classes of memory that do not have this built-in safeguard. Among these are memories for events, even for quite recent events. A professor of criminal law had just started an introductory lecture on evidence to a large class of students when a villainous-looking young man, in blue jeans and a black leather anorak, burst into the class brandishing an automatic pistol. The intruder shouted menacingly and abusively at the professor, fired two shots at him which missed, pulled off the professor's jacket, grabbed his wallet from an inside pocket, fired a third shot and ran out of the room. The professor showed admirable composure and the students were deeply impressed when he picked up and dusted off his jacket, put it on, asked for silence, and continued with the lecture. Near the end of the period the professor summed up, then said that he wished to call for evidence about what had happened at the start of the lecture. He then questioned many of the students and was informed that the intruder had been a young or middle-aged man, or a young woman, with long or short, straight or wavy hair, that was black or brown or nearly blonde. He had been wearing a black or brown or blue anorak or coat over blue or black jeans and was carrying a revolver, or automatic pistol or sawn-off shotgun, and had fired one, two, three, four or five shots. He had stolen a wallet or . . . and so on.

It was an impressive demonstration of the variability of observation and of memory over a period of less than an hour and was, of course, a deliberate put-up job using an actor, designed to induce in the students a healthy scepticism about the reliability of evidence. Recollection of events can be of the highest importance especially if given as evidence of wrong-doing in criminal cases. A good witness will usually be able to give reliable and acceptable evidence of the salient points of an event, but even the best of witnesses is likely to be inaccurate on many minor details.

In cases of alleged childhood sexual abuse, the entire case may depend on the uncorroborated evidence of a person's apparent early memories. Just how unreliable these memories can be has been shown by research carried out in recent years. This is vital research because whole lives are ruined by successful indictments for sexual abuse. In this context, the research has shown that defects far more serious than that of inaccuracy on detail can occur. In many cases, the memories may be entirely fictitious.

Inevitably, the subject of childhood sexual abuse arouses strong emotions and such emotions are apt to cloud judgement. There will often be a strong bias in favour of the alleged victim and against the alleged abuser. The way in which the memories of abuse are elicited appears to be very important if the production of false memories is to be avoided. One therapist who has written a popular book on the subject recommends that the alleged victim should be urged to spend time imagining that he or she has been sexually abused. The 'victim' should be encouraged to 'give free rein to the imagination', to forget about accuracy but to ask herself or himself questions such as 'Where are you?' 'Is it indoors or outdoors?' 'What time is it?' and even 'Who would have been most likely to do it?'

Research carried out by Elizabeth Loftus, professor of psychology at Washington University, has shown that it is just exactly this kind of suggestion that is liable to plant entirely false memories of an event in the mind of an alleged victim. Professor Loftus is a specialist in memory, evidence and courtroom procedure. Her book *Eyewitness Testimony* won a National Media Award from the American Psychological Foundation. She refers to cases in which psychotherapists have effectively fed misinformation to their patients that have led to unfounded public accusations by them of earlier sexual abuse by parents. This has produced ruinous damage to innocent people which cannot be undone even when clear medical evidence can be produced that no such abuse could have occurred.

Research has shown that it is all too easy to induce false memories, even to persuade a person that he or she remembers events that occurred in the first few weeks of life at a time when the part of the brain responsible for the registration of memories (the amygdaloid nucleus) has not yet developed to the stage at which such registration is possible. In Professor Loftus's words: 'False memories are constructed by combining actual memories with the content of suggestions received from others. During the process, individuals may forget the source of the information.'

Since Freudian psychoanalysis is based so largely on sex, this work has strong relevance to it. An analyst who is trained and conditioned to be constantly looking for sexual clues in the memories of his or her patients would hardly be human if he or she did not, from time to time, unwittingly plant false memories. If these 'memories' are then used to build up a body of 'fact' to account for this or that psychological problem, the procedure would seem to be likely to do a great deal more harm than good. At best, it would seem unlikely to be particularly therapeutic. It is not beyond the bounds of possibility that the founder of psychoanalysis may have done this, and that the foundations of the discipline may be based on nothing more substantial than induced memory.

In October 1997, the British Royal College of Psychiatrists agreed to the publication of a report which advised its members not to use forceful or persuasive interviewing methods, or the so-called 'memory recovery techniques' to unearth 'memories' of sexual abuse. The College authorities were unwilling to publish the report themselves, which had been produced by a working party appointed by the College. This was because of the controversial nature of the subject and because of the need to try to protect the rights both of abused children and of people accused, as a result of such memories, of sexual abuse.

The 'false memory syndrome' has aroused a great deal of argument and dissension in psychiatric circles both in UK and the USA. But the committee concluded that false memories *can* be created and the report provides clear guidelines as to how psychiatrists should avoid causing these. Certain practices were condemned outright. As in all such matters, a balanced view must be taken. To argue that recovered memories were either all true or all false was clearly a serious mistake. The chairman of the committee, Professor Sydney Brandon, said that false memories were more likely to be created when the therapist 'believes that all adult relationship problems are related to childhood memories of abuse.'

Technology

> The paper is nothing but nonsense, unfit even for reading before the Society.
>
> *A referee of the Royal Society reporting on an 1845 paper by the Scottish civil engineer and physicist J. J. Waterston which developed the foundations of the kinetic theory of gases and thermodynamics, and anticipated the later work of Clausius, Joule and Maxwell*

The Tay Bridge disaster

The Tay is the longest river in Scotland. It arises on the slopes of Ben Lui in the Grampians and flows 120 miles to the North Sea below Dundee. The wide estuary of the Tay had long been a great inconvenience to travellers on the east coast of Scotland and, with the growth of the railways in the first half of the nineteenth century, this inconvenience became intolerable. In 1863 the first official proposals for a bridge over the Tay were made public and in 1870 an Act of Parliament was passed authorizing the construction of an iron railway bridge over the river.

The engineer in charge of the project was Thomas Bouch (1822–

80). Bouch was a highly successful civil engineer who already had to his credit some notable engineering works. He had been heavily involved in the construction of the Lancaster and Carlisle railway; he had been resident engineer on the Stockton and Darlington railway and, from 1849, manager and engineer of the Edinburgh and Northern railway; he had developed a 'floating railway' system for shipping goods trains across the Tay and Fort estuaries; and he had designed nearly 300 miles of railways in the north of England and Scotland, including the South Durham and Lancashire Union.

Bouch was especially good at bridges and had been responsible for the construction of many, especially railway bridges. The Tay Bridge, however, was by far his greatest project. The bridge, which was just under two miles long, crossed from Newport to Dundee. There were 85 spans of wrought-iron lattice girders, 13 of which were over the deep navigable water. The work was completed on 22 September 1877, and was passed by a Board of Trade inspector, Major-General Coote Synge Hutchinson, late of the Royal Engineers. The bridge was opened on 31 May 1878.

It was the acme of Bouch's career. He was given the freedom of Dundee, and on 26 June 1879 he was knighted. The celebrated Scottish poet William McGonagall (1830–1902), wrote a remarkable ode on the occasion of the opening. It includes the immortal lines:

> *Beautiful Railway Bridge of the Silvery Tay!*
> *And prosperity to Messrs Bouch and Grothe*
> *The famous engineers of the present day,*
> *Who have succeeded in erecting the Railway*
> *Bridge of the Silvery Tay.*

All went well until the evening of Sunday, 28 December 1879, when it became apparent that Sir Thomas had blundered. In fairness, it must be said that, at the time, not much was known scientifically about the effects of wind pressure on structures of this kind. In fact the engineers were using figures that greatly underestimated the matter. A hundred years earlier, in 1759 the civil engineer John Smeaton (1724–92) had presented a table to the Royal Society and this was the basis of the engineers' calculation. Smeaton was interested in wind and wrote a book called *The Natural Powers of Wind and Water to Turn Mills*. His table stated that 'high winds' exerted a pressure of 6 pounds per square foot; 'very high winds' exerted pressure of 8–9 pounds per square foot; and 'storms or tempests' could exert as much as 12 pounds per square foot.

Unfortunately for Sir Thomas, his faith in this received wisdom was unjustified. He had provided no continuous lateral wind bracing below the deck, and, in fact, there is evidence that he had made no special provision whatsoever to combat the effects of wind pressure. On the fatal Sunday evening the elements refused to conform to Smeaton's rules and exposed the new bridge to a violent hurricane. As ill-luck would have it, the Edinburgh mail train carrying 70 passengers was crossing the central spans when the force of the wind blew 13 spans into the river carrying the whole train with them. All the passengers died.

Sir Thomas was not long in following them. We must suppose that the stress of this terrible reversal in his fortunes affected his health. He had produced a design for an equally impressive project – a suspension bridge over the Firth of Forth. But as a result of the Tay disaster public confidence in him was severely shaken. The four railway companies that were promoting the construction of the Forth suspension bridge abandoned the project and tore up Sir Thomas's plans. The unlucky engineer died on 30 October 1880.

A report of the Court of Inquiry into the Tay Bridge disaster was published in a Parliamentary Paper in 1880. As a result of this it was recommended that, in the case of the Forth Bridge, the plans should allow for a wind pressure of 56 pounds per square foot. This was gross overreaction, based on emotion rather than reason and even at the time the figure was criticized by distinguished engineers who had carried out wind-pressure tests, using gauges installed at the site, over a period of two years. Later research showed that 34 pounds per square foot would have been more than amply safe.

> This is the biggest fool thing we have ever done. The bomb will never go off, and I speak as an expert in explosives.
>
> *Admiral William D. Leahy (1875–1959), advising*
> *President Truman on the atom bomb in 1945*

The unnecessary atomic bomb

Hitler's aggressive behaviour in 1938 – his seemingly confident challenge to Britain, America and the other allies – was not only seriously worrying; it also led many scientists to ask themselves whether there was something more behind it than just his superiority in tanks and aircraft. There was plenty of talk of a secret weapon.

Those who were most worried were the nuclear physicists, because they were well aware that the possibility existed that German scientists could be working on an atomic bomb. This possibility was consistent with Hitler's seemingly reckless disregard of the limitation in Germany's sources of raw material and its limited manufacturing potential compared with that of the rest of the free world.

By 1938 those workers in Britain and America who were engaged in nuclear research had already cut off relations with the German scientists. Science which, only a few years before, had been completely open, with immediate publication to the world of all new discoveries, had closed in. Four years earlier, at the 1934 Würtzburg Congress of Physics, there had been violent arguments between a clique of 'German' physicists and those who held that race and nationality had nothing to do with science. The chauvinism of the Germans had shocked the others and it was clear that there was something political behind it.

By early 1939 American physicists were in almost unanimous agreement that there should be no further scientific communication with physicists working in Nazi Germany. All were well aware of Hitler's ruthlessness. They knew that he was perfectly capable of exerting extreme pressure on scientists – even threats of imprisonment of their families – to direct their efforts to the development of weapons. This was a double-edged decision. Clearly it was essential not to pass on new knowledge that could further the war effort of a potential enemy; but at the same time, insulation from the German scientists meant that the Allies were no longer aware of what was going on in Germany.

In April, however, some of the scientists had received confidential information that strongly suggested that the Germans were about to start working on an atom bomb. Two German physicists had informed the head of the Department of Research at the German Ministry of Science that a 'uranium machine' was a possibility. This caused the official to call an immediate meeting of six German atomic physicists. The meeting was held in an office in Unter den Linden, Berlin.

Those present were instructed to consider everything discussed as secret. One scientist disobeyed and revealed that the meeting had been about the possibility of nuclear-powered vehicles, not atom bombs. The result was that a German physicist, Dr S. Flügge, who strongly believed that a discovery as important as nuclear fission should not be kept secret, decided to publish an account of the uranium chain reaction in the journal *Naturwissenschaften*.

When his article appeared, the American and British scientists were horrified. The appearance of such an article meant that its publication had been approved by the Nazi authorities. This clearly implied that they believed themselves to be well ahead of the Allies in atomic research. By mid-1939 the American scientists were convinced that they had to persuade the US government of the risk from Germany. Niels Bohr (1885–1962) spoke at a meeting of the American Physical Society and warned that a bomb containing a tiny amount of uranium-235 (see The breeder reactor blunder) could destroy a whole city.

Werner Heisenberg (1901–76) visited the United States in mid-summer 1939 and Columbia University tried to persuade him to stay by the offer of a professorship, but he refused. He said he was sure that Hitler would lose the war but he wanted to remain in Germany to help in the ensuing chaos. In conversation, much later, with Enrico Fermi (1901–54), the Italian-born US nuclear physicist who had built the first nuclear reactor, Heisenberg pointed out that, in the summer of 1939, there were 12 nuclear physicists who, by getting together and making a firm agreement, could have prevented the atomic bomb from ever being made. Tragically, by then scientists were too divided, and it was questionable whether, as a group, they would have possessed enough political imagination and moral courage.

News reached the United States that there had been a second meeting of nuclear scientists. This one was called by the German Army Weapons Research Department and it had been called to discuss the question of starting a chain reaction in uranium. At the same time the Americans learned that the Germans, who now occupied Czechoslovakia, had published orders forbidding the export of uranium from that country. Apart from Belgium, Czechoslovakia was the only country in Europe with substantial stocks of uranium.

Even now, the scientists were having a very hard time trying to persuade the US government that they should take the threat of a German atom bomb seriously. Earlier in the year Fermi had had a meeting with an Admiral Hooper, Director of Naval Operations Technical Division, but the Admiral had been unimpressed. Even Bohr's statement at the American Physical Society meeting had done nothing to alert either the US service chiefs or the government.

The news about the veto on uranium export suggested that Hitler might quickly go for the Belgian stocks, so the scientists went to see Einstein, who was a friend of the Belgian Queen Mother. The Belgians got their uranium from the Congo. It was hoped that Einstein would agree to warn the Belgians in this way. Einstein was willing to do more than this. He would sign a letter to the President of the United States urging him not only to advise the Belgian government to safeguard its uranium stocks, but also to agree to the allocation of funds for the acceleration of American research into the production of an atomic bomb.

Einstein had not been involved in nuclear research and knew little about it, but his name was known to everyone and he was an immensely respected figure. As soon as he heard about the possibility of a neutron chain reaction, he got the message and agreed to sign a letter drafted by the physicists. This was duly delivered to the President and the result was the Manhattan Project.

At this stage no one could be certain that an atomic bomb could be made. There were enormous problems. There was no questioning the energy that could be produced. A kilogram of matter converted entirely into energy would supply the entire energy needs of the whole of Britain for a year. The question was whether some of this energy could be released as an explosion. First it was necessary to purify enough of the fissionable uranium-235 (U-235) and this was extremely difficult. U-235 is an isotope of the much commoner U-238. This means that it has exactly the same chemical properties as U-238 so can't be separated by chemical means. The only difference between the two isotopes is that one has three fewer neutrons in the nucleus than the other making the atom ever so slightly lighter.

The scientists tried separation by the principle used in mass spectroscopy. This involved exposing the U-238 in gaseous form to powerful magnetic and electric fields, causing the atoms to move. The lighter atoms were able to move slightly faster and so could be separated. The method was, however, horribly expensive of energy and produced only tiny amounts of U-235. One prototype equipment, working flat out, produced only 1/30,000,000 of an ounce in a day. The equipment also required heavy electrical conductors and, as copper was scarce, the US government agreed that Treasury silver worth $400,000,000 could be used.

The separation method finally used was, however, the gaseous diffusion method, requiring a plant that covered half a square mile. This method depends on the fact that the speed of movement of the molecules of a gas is inversely proportional to the square root of the weight of the molecules. So if two hot gases, such as gaseous compounds of U-238 and U-235 are allowed to diffuse through tiny holes there will, initially, be a slightly higher concentration of U-235. If this process is then repeated many times, there will be a progressive rise in the concentration of the lighter gas.

The only suitable compound turned out to be uranium hexafluoride – a horribly poisonous and corrosive substance. This becomes gaseous at a temperature of 74.4°C. The difference in the molecular weights of the two isotopes is very small and the difference in the square roots is smaller still. It was found that the best enrichment they could achieve in a practical system was 30 parts per million. So about 5000 consecutive diffusion filters covered with tiny holes were needed, and these had to be made of a material that would resist the corrosive effect of the gas. The holes had to be less than a two-millionth of an inch in diameter and had to be clog-resistant. Billions of holes were necessary. The pumps that moved the gas also had to be capable of resisting corrosion and the lubricants used had to be able to resist forming chemical compounds with the highly reactive uranium hexafluoride.

Then there were the problems of ensuring that the bomb would work as a bomb. No one knew precisely the critical size of a U-235 bomb. This is the amount of uranium that, when brought together into one lump, will undergo an explosive chain reaction. If the size was too small the bomb would not explode; if too large it would explode prematurely, probably in the production plant. It was also essential that the safe, below-critical, pieces should be brought together very quickly indeed. In a critical mass the chain reaction is so fast that the explosion occurs in about a millionth of a second. Obviously, the only way the parts could be brought together fast enough was by using a conventional explosive to blow the subcritical parts together.

After all this incredible expense and effort, the U-235 derived from the enormous diffusion plant was never used to make a bomb. In parallel with this method, another process involving what is called a breeder reactor had been set up at Hanford, Washington. This was really a kind of atomic power station and its product was another fissionable element called plutonium. It was this that was used in the two bombs. But there is another, possibly even more important, story to be told about plutonium and you can read about this in The breeder reactor blunder.

By the time the first bomb was ready for trial, the whole project had cost two billion dollars. Never before had so much sustained effort and expense been devoted to the ends of sheer destruction, and the explosion of the bombs over Hiroshima and Nagasaki wonderfully concentrated the minds of those responsible. Among many others, Albert Einstein was later greatly to regret his impetuosity. At the time, he felt he had no choice. But later, in retrospect, when it became clear that the Germans had had no possible chance of making an atom bomb, he was bitterly distressed. Post-war research revealed that Hitler had hated the scientists and had thoroughly antagonized them. Many of the best of them had been driven to escape from Germany out of fear of Nazi persecution. Others were simply called up to fight on the various battle fronts. Hitler's attitude to science meant that very little of the limited resources of the Reich were devoted to scientific research. The net result of this all this was that few if any of the German physicists had the least motivation to press for research into the production of an atomic bomb. Indeed, mercifully, the scientists were remarkably successful in leading the Army authorities attention away from the idea.

Ironically, nearly all the uranium oxide in Germany had been bought up by the German Army Ordnance Department, not to use for atom bombs but to try to make an alloy for armour-piercing shells. Uranium is a very heavy metal and the metallurgists thought it could be effective in this role. This was a hopeless and dangerous idea which, of course, failed. But the Ordnance refused to release the uranium to anyone else.

Notwithstanding all these difficulties, a German Uranium Project was actually started. Werner Heisenberg was in charge and he insisted later that he had only pretended to collaborate with the Nazis. There is a strong implication that he believed that by retaining control of nuclear research he would be able to prevent some other physicists from actually trying to make a bomb for Hitler. Significantly, as early as the end of 1939, Heisenberg had worked out the theoretical basis of an atom bomb. He and a very few of his closest and most trusted associates, however, kept this information to themselves. Their official report, however, was: 'We are aware of no practical way, during the war, of producing an atomic bomb with the available resources in Germany. We must, however, continue with our research to ensure that the Americans also will be unable to make a bomb.'

Heisenberg and his friends felt it important to ensure that the authorities remained sufficiently interested to continue to authorize the release of young physicists from military service. Heisenberg was much criticized for the stance he took. Even some of the German scientists blamed him for seeming to cooperate with the Nazis.

> These so-called cosmic rays certainly don't come from the cosmos.
>
> *William Francis Swann,*
> *Director of the Franklin Institute of Physics*

The Challenger space shuttle failure

The year 1986 was to be the 'Year for Space Science'. NASA was riding high and was coming to the fruition of a number of major space projects that had long been under development. At the heart of the great programme was the space shuttle – a magnificent, reusable space vehicle that could carry seven people and an enormous payload. The shuttle was the first spacecraft to land on earth.

Known originally as the Space Transportation System (STS), the shuttle first flew in April 1981. It is a massive vehicle, and massively expensive, but designed to be reused about a hundred times. It is carried into orbit by rockets and returns and lands like a normal aircraft. It has a payload of almost 30,000 kilograms (65,000 pounds) and can bring back a substantial payload from space to earth. The shuttle, also known as the 'orbiter', is carried into orbit by rockets. Three of these are liquid fuel rocket engines on the orbiter itself and two others are solid fuel rockets. Lying between the two solid fuel engines and underneath the orbiter is an enormous fuel tank for the shuttle engines. This tank, which dwarfs the vehicle, contains, at the time of lift-off, 703,000 kilograms (1,550,000 pounds) of propellant. This consists of liquid hydrogen and liquid oxygen contained in two separate pressurized compartments – an almost unthinkable amount of chemical energy locked up in about the most concentrated form imaginable.

Fuel from the vast external tank passes to the shuttle to supply its reusable engines. These engines are controllable by the pilot and can be throttled to provide a range of thrust which is maximal at take-off and is then reduced so as to limit the g forces on the crew to about three times that of gravity on the final stages of the ascent. The engines are mounted on movable gimbals for purposes of steering and for correcting pitch, roll and yaw movements. Just before the vehicle reaches its orbital velocity, the external tank, having served its purpose, is disconnected and jettisoned at a point over the ocean where it can safely be left to break up and burn as it re-enters the atmosphere.

The system would not be able to take off were it not for the two massive solid fuel booster rockets. Each of these produces a thrust of over 11 million newtons (1,600,000 foot pounds) and the two are ignited at the same time as the three liquid fuel engines on the shuttle. When they have served their purpose in producing the required take-off thrust and orbital velocity, they are disconnected and freed. Unlike the main fuel tank, they are, however, fitted with parachutes and are reclaimed to be used again. Each booster rocket has four segments and between each of these segments are two rubber ordnance-rings (O-rings) to seal the joint.

Rocket engines do not provide propulsion by pushing back against the air, as some people think. If this were so, they would not be able to work in space, where there is no air to push against. They work by producing a thrust against the inside of the engine opposite to the open nozzle. Enormous pressures are generated inside the engines as the fuels, whether liquid or solid are rapidly converted to very large volumes of gases. If there were no outlet from the engine, the pressure would be exerted equally on all the internal walls of the engine. But because the bottom end of the engine is open there can be no pressure at this point. The pressure on the side walls is equal all round and balances out. But the pressure on the wall opposite the open nozzle is not balanced and so it drives the rocket upwards.

Mission 51-L was designed to take into orbit the second tracking and data relay satellite and a smaller satellite that was to conduct observations of Halley's comet as it neared the sun and then be retrieved after two days. An unusual feature of this mission was that the shuttle would carry a non-technical member, the school-teacher Sharon Christa McAuliffe, who had secured her place on the trip by winning the national screening programme that had started two years before. Sharon was not the only woman on board. The other members of the crew were Francis R. Scobee (mission Commander), Michael J. Smith (pilot), Ellison S. Onizuka (mission specialist), Judith A. Resnik (mission specialist), Ronald E. McNair (mission specialist) and Gregory B. Jarvis (payload specialist from Hughes Aircraft company).

The flight had to be postponed four times prior to the launch at 11.38 a.m. on 28 January. On that day the weather was very cold.

Less than a second after the ignition of the solid booster rockets puffs of grey smoke emerged from the rear field joint on the right hand booster. This smoke quickly turned black. The appearances indicated that the joint was not properly sealed and that the rubber O-ring seals were being burned away by the hot gases within the rocket. This event was not seen by observers on the ground and was revealed only later when films were analysed. The leak at the seals was quickly closed by residue from the burning solid fuel and the craft took off. Fifty-eight seconds later the leak opened up again and an intense, 8-foot flame like a blow-torch shot out. This quickly increased to 40 feet long and was driven back in the slipstream against the side of the external fuel tank. Fourteen seconds after the flame appeared the lower strut attaching the booster rocket to the fuel tank was burned away and broke loose. The rocket swung round on its upper strut and its top smashed into the main fuel tank, which came apart. At that point the whole shuttle vehicle broke into pieces. One piece was the crew compartment which continued to coast upwards for a time and then plunged 47,000 feet into the Atlantic. The crew were not wearing space suits and the immediate loss of cabin pressure would have rendered them unconscious within seconds. Fuel from the tank escaped and ignited to form an enormous fireball. The crew compartment was never found.

The word 'blunder' is hardly appropriate.

Many factors contributed to this disaster, but the immediate cause was quickly established. The rubber O-rings had been hardened by the extreme cold and had lost their resilience. This was demonstrated by the scientist Richard Feynman at one of the enquiry meetings using a glass of ice water at 0°C, a piece of the O-ring and a pair of pliers. When the three shuttle engines fire the vehicle is momentarily pressed forwards and then swings back. The booster rockets are then ignited. In this case the hardening of the rings had allowed the joint to open a little and rotate but had prevented resealing. So the joint was left open and when the rocket fuel started to burn the flame was able to get through the gap. That was what killed these seven courageous young people.

The enquiries found that several engineers had, for two years, been expressing concern about the reliability of the booster rocket seals. Engineers had even warned their superiors about the danger the night before this particular launch. Behind this technological factor was a more fundamental cause. NASA was trying to do too much with resources that were too limited. Once the schedule had been decided, senior managers had pressures placed on them to push on with the programme – pressures that could lead to dangerous decisions.

Richard Feynman was later to comment: 'For a successful technology, reality must take precedence over public relations, for Nature cannot be fooled.'

In the papers of Thomas Young I can find nothing which deserves the name either of experiment or discovery. I deem them destitute of every species of merit. The Royal Society is to be censured for printing such paltry and unsubstantial papers.

Lord Henry Peter Brougham FRS (1778–1868),
writing of the originator of the wave theory of light
and many other principles and ideas now
known to be of fundamental scientific importance

Chernobyl

On 28 April 1986 Swedish monitoring stations suddenly realized that abnormally high levels of wind-transported radioactive dust were coming from the Soviet Union. They immediately demanded an explanation. At first there was an attempt at a cover-up, but soon the Soviet government admitted that there had been an accident at its Chernobyl station, and the attention of the world was directed to this previously obscure spot.

The Chernobyl nuclear power station is situated in the Ukraine at Pripyat, 10 miles from the city of Chernobyl and 65 miles from Kiev. By the time it was fully commissioned in 1983, there were four water-cooled, graphite-moderated reactors, each of which produced enough power to generate a million watts of electricity. There are many different designs of nuclear reactors which vary in the way in which the chain reaction in the core is moderated and the rise in temperature kept within safe limits. Several different types of coolants are possible, including plain water, heavy water and even molten metals with a low melting point. The water-cooled graphite-moderated design of reactor had been considered by Western experts but had been rejected.

Two years after it went on-line, the authorities at Chernobyl decided to run a test on number 4 reactor to find out how long the turbines would continue to turn the dynamos to generate power after the reactor was suddenly switched off. This was important because the electric power for the emergency safety systems was provided by the station's own turbine-operated generators. The turbines would not stop generating suddenly but would gradually run down after the steam supply had been cut off. The technicians wanted to know how long electricity would be available.

So, on 25 April 1986, they started the test. The first step was to reduce the reactor to half of full power. Thirteen hours later they disconnected the emergency core cooling system. It was now 2 p.m. and at this point the next stage of the test had to be postponed because the electric grid had to be supplied with power. So the reactor operated for the next nine hours with the emergency core cooling system nonoperative. When the grid demand dropped again the power of the reactor was further reduced and the computer that provided automatic operation of some of the control rods was switched off.

The test was now resumed and, with some difficulty, the reactor output was reduced to only 7 per cent of full power. By now, the technical staff had shut down the emergency water-cooling system, switched off the emergency shutdown system and the power-regulating system, and had withdrawn nearly all the control rods from the reactor core. The operators now shut off the steam supply to the turbines. This proved to be disastrous. As the turbines slowed down, the electricity output dropped and the pumps circulating water to cool the reactor core also slowed down. The result was a sharp rise in the temperature of the core, and, because all the automatic safety controls had been turned off, the reactor quickly went out of control.

As soon as this was apparent, an emergency order was given to push in all the control rods and shut down the reactor. But by now the core was so overheated that most of the control rods would not go right in. For want of moderation the chain reaction in the reactor core speeded up. A critical point was quickly reached at which the reactor was more like a bomb than a controlled fission device. In three seconds the output power rose from 7 per cent to 50 per cent of full power. The nuclear fuel elements burst so that the very hot fuel came directly into contact with the cooling water which immediately turned to superheated steam. This steam reacted with the graphite and the zirconium coating of the fuel rods to generate large quantities of hydrogen – a highly inflammable gas.

At 1.23 a.m. on 26 April there were several explosions. An immense fireball formed which blew off the heavy protective steel and concrete lid of the reactor and the roof of the building. The massive servicing crane that overtopped the building crashed down onto the reactor. Two people were killed at once and 31 received a lethal dose of radiation from which they died within a few months. These people suffered severe skin burns and destructive radiation effects on their most rapidly reproducing tissues such as the lining of their intestines. Many of them died from total failure of the blood cells producing tissue in the bone marrow. Unhappily, bone marrow transplantation proved to be almost totally unsuccessful.

A very large amount of radioactive material – some 8 tons – was forced up into the atmosphere. This was a greater quantity than was released by the two war-time atomic bombs at Hiroshima and Naga-saki. About half of this dangerous material rose to about 20,000 feet and drifted eastward. The remainder went north-west at low altitudes. Some of it was carried as far west as France and the west coast of Ireland. Some even got as far as north America. The whole of Europe except parts of Spain was irradiated by fallout. To the south, the fallout radiation reached Tunisia and other north African countries. To the east, radioactive material reached Israel and Japan.

On 27 April the evacuation of the population of Pripyat – some 30,000 people – began. A number of heroic workers remained on the station to try to contain the problem. For 20 miles around the site of the disaster the ground and the water supplies were heavily con-taminated by radioactivity. In an area of 300 square miles all the inhabitants had to be evacuated. As a result of the radiation received, some 200 people developed severe radiation sickness and some of these died. Many farm animals were irradiated and for years after-wards a large number were born deformed.

The Soviet authorities did all they could to underplay the disaster. Initially, they reported that the levels of radiation around the power station did not exceed 15 milliroentgens, an almost insignificant level. In fact they were at least a hundred times higher than this. This is, of course, residual radiation. At the time of the explosion, people in the vicinity must have received well over a thousand times this dose. The lethal short-term dose of radiation is around 500 roentgens. Experts have calculated that the additional radiation, worldwide caused by Chernobyl, would, over a period of 50 years, cause an additional 7000 deaths from cancer. This may seem a substantial figure, but it must be put in perspective against the total of deaths from cancer in this period – about 100 million.

By November 1986, reactor 4 at Chernobyl was finally sealed. In July of the following year, six people deemed responsible for the disaster were brought to trial, convicted and sentenced to terms of imprisonment of from two to 10 years.

He was one of the most clear-sighted men who have ever lived, but he had the misfortune to be too greatly superior in sagacity to his contemporaries. They gazed on him with astonishment, but could not always follow the bold flights of his intellect, and thus a multitude of his most important ideas lay buried and forgotten in the great tomes of the Royal Society of London, till a later generation in tardy advance re-made his discoveries and convinced itself of the accuracy and force of his inferences.
Hermann von Helmholz, writing of the great English physicist, physician and Egyptologist Thomas Young (1773–1829)

The wrong-spot moon landing

The NASA Apollo Project to put humans on the moon was started in 1961 and the first flight of the project was in October 1968. Between then and May 1969 there were four flights, *Apollo 7* to *Apollo 10*. All of them orbited the moon. *Apollo 8*, from 21 to 27 December 1968 was the first manned orbit around the moon. The craft took up an orbit about 70 miles (112 kilometres) above the surface of the moon and remained in it for about 20 hours, transmitting TV pictures back to Earth. At that stage, the information obtained suggested that lunar landmarks could be used for navigation to lunar landing sites.

Apollo 11, with Neil Armstrong, Edwin E. Aldrin, Jr, and Michael Collins on board, was scheduled to land two humans on the moon on 20 July 1969. Michael Collins would remain aloft controlling the lunar orbiting module. The selected landing zone was in the Sea of Tranquillity. This is not, of course, an actual sea, but these many circular areas with raised margins had been called 'seas' by early astronomers. This area is very large and is by no means all tranquil, so a smooth spot had been picked out. Unfortunately, putting the landing module Eagle down on that spot turned out to be far more difficult than most people had anticipated. So difficult, in fact, that, in the end, the lunar module actually landed in the wrong place. This made it, in retrospect, one of the most conspicuous, if unavoidable, errors in the history of science.

The NASA engineers had been having problems in getting their computer software right. The main, and quite unexpected, reason for this was that the moon's gravitation was found to vary markedly over its surface. This was causing orbiting lunar modules to vary in speed. The engineers had discovered from studying the orbits of previous modules that this variation was due to the moon's seas. Every time a vehicle in orbit passed over one of the seas, it accelerated slightly due to a slightly stronger gravitational pull. Then, when it had passed, it slowed down again. The effect, however, was to change, in seemingly unpredictable ways, the time it took to fly round the moon.

This increased gravitational effect had to be due to increased density in or around the seas. The reason for this remained speculative, but it was probably the result either of asteroids that had hit the surface or of the upthrust of more dense material from the moon's core. The cause was not the problem. The problem was that the engineers now had a horribly complex task in trying to write software that would take into account all the numerous perturbations in the gravitational pull experienced by the module as it moved round the moon. Software that realistically represented all the perturbations had to contain so many mathematical terms that it was too big to be used by the very limited small on-board computers of the time. Simpler programs using less memory would be necessary.

This was just the start of it. When they tried to check how accurate the results of their navigational programs were, they found that it was virtually impossible. Repeated observations of moon surface features were made from orbiting craft. For these to be navigationally useful, it was necessary to know the position of the spacecraft at the moment of observation. Since this was what the exercise was all about, it was not particularly surprising that they found that the apparent position of moon features varied by as much as several kilometres, even when the observation was repeated during different orbits on the same mission.

Yet another problem was the fact that orbiting vehicles could be tracked from Houston only when they appeared from around the back side of the moon. So the data on the orbits contained no information about what actually happened when the vehicles were out of sight. The best that could be done was to compare the actual time of appearance with the expected time, thus obtaining an error that was the sum of all the errors occurring during the transit behind the moon. These problems compounded each other to such an extent that, in the end, the engineers ran out of time.

Schedules had to be met. Political and other pressures made it impossible to postpone the much-heralded landing of men on the moon. It was decided that, for *Apollo 11*, each time the lunar orbiting module appeared from round the back of the moon, the astronauts on board would be supplied verbally by the Houston staff with information to correct the flight path and speed. The trouble with this method was that the supply of fuel on the lunar module was very limited and each correction used up more of it than was desirable. There was, however, no practical alternative.

When Neil Armstrong, who was manually guiding the lunar landing module, and Buzz Aldrin, who was with him, first caught sight of the expected smooth landing site, they were, no doubt, more than mildly annoyed to see that it was covered with boulders. At an altitude of about 300 metres the module was, apparently, heading for disaster. At 100 metres altitude the surface was pockmarked with craters and boulders. With commendable calm, Armstrong steered the craft around a field of rocks that could hardly be less suitable for a landing. The dilemma for the Mission Control staff at Houston must also have been excessively painful. It was a very tense moment. The whole world was watching on TV. Everyone, from the President of the United States down, was expecting a landing. There was only a minute's fuel left before the landing would have to be abandoned and Armstrong ordered to turn on the ascent engine – which had its own fuel supply – and return to the orbiting command module above them. This would have been the ultimate embarrassment.

Almost at the last moment, Armstrong found an acceptably flat landing site. The rest is history. Later analysis showed that the module had touched down on a plain near the south-western edge of the Sea of Tranquillity, about 6 kilometres past the intended point and about 2 kilometres off course to the south.

> However fascinating it may be as a scholarly achievement, there is virtually nothing that has come from molecular biology that can be of any value to human living.
>
> *Frank MacFarlane Burnet (1899–1985),*
> *Nobel Prizewinning immunologist whose*
> *work made organ transplantation possible*

The tragedy at Bhopal

Bhopal city is the capital of the state of Madhya Pradesh in central India. It has a population of over 670,000 people and has the largest mosque in India. On 3 December 1984, the worst industrial accident in all history occurred there.

In the small hours of the morning, at an insecticide plant owned by a subsidiary of the American Union Carbide Corporation, a large volume of water entered a tank of the highly toxic compound methyl isocyanate. This compound reacts violently with water, and about 45 tons of methyl isocyanate gas escaped and drifted in a cloud over the densely populated city. The gas, denser than air, blew slowly over the town and descended on the inhabitants.

At first there was confusion as to the nature of the gas that had escaped. The doctors, trying to help the enormous numbers of casualties, were uncertain. One senior doctor said that people were showing signs of cyanide poisoning; another disagreed. Some 2,500 people were poisoned and died within 12 to 72 hours. The doctors did not know what was best to be done. A telex was received from the Union Carbide Corporation headquarters advising that if people had symptoms of cyanide poisoning they should have sodium thiosulphate and amyl nitrate injections. This advice was not passed on.

Most died from a condition known as pulmonary oedema in which the lungs fill up with fluid. About 90,000 people suffered lung and eye irritation sufficient to require medical attention, and many of these became permanent invalids. One of the worst long-term effects was bronchiolitis obliterans, a condition of blockage of the smaller air tubes in the lungs. No one knows how many people have since suffered and died from this persistent condition.

Methyl isocyanate or methyl mustard oil is a highly toxic compound of cyanide used as an agricultural pesticide. It is intensely irritating to the eyes and to the nose, throat and lungs. The effects are intense weeping, sore throat, a choking sensation, coughing and vomiting. The compound is very damaging to the outer lens of the eye (the cornea) and leads to painful ulceration followed by opacification. It is estimated that some 50,000 people were blinded, at least temporarily, at Bhopal. Many more suffered permanent visual loss, some from the direct effect of the compound and some from secondary infection and further corneal damage.

The lung damage also opens the way to secondary infection such as pneumonia and this caused many of the deaths. Methyl isocyanate also causes a rise in the spontaneous abortion rate in pregnant women and a significant rise in the mortality rate in newly born babies. The toll of human suffering occasioned by this catastrophe is hard to imagine. Many thousands of people who survived the occasion are suffering from it still.

Investigations made after the Bhopal disaster showed that the plant was seriously understaffed and that seriously low standards of industrial safety were to blame. The investigators found 65 management errors, 12 operator errors, 21 equipment failures and 28 breaches of regulations. One of the directors of the company suggested that the release of the gas had been due to sabotage, but this was ruled out. The legal arguments about the matter went on for years. The Indian government wanted injury claims to be heard in the USA. An American Federal judge ruled against this and the cases were heard in New Delhi. The matter was soon adjourned, however, because the Union Carbide Corporation was not represented.

In February 1989, the Union Carbide Corporation finally settled with the Indian government. By then 3,500 people had died as a result of the release of gas. The company agreed to pay $470 million in settlement of all claims against it. All criminal charges against company employees were dropped.

> I have not the smallest molecule of faith in aerial navigation other than ballooning, or of the expectation of good results from any of the trials we hear of.
>
> *Lord Kelvin (1824–1907), writing to Baden-Powell in 1896*

The Hubble telescope – an expensive blunder

Since the time of Galileo astronomers have been persistently annoyed by the effect of atmospheric conditions on their telescopic observations. Irregular air movement, atmospheric dust and gases, light pollution and other factors have all contributed to the degradation of the images produced by terrestrial telescopes. This is why so many observatories have been built in remote places and on high mountains. In addition, the atmosphere severely attenuates the amount of ultraviolet light reaching the earth. This is good news for the doctors but bad news for the astronomers who are interested in getting the widest possible range of radiation in the area of the light spectrum. So, for centuries, it has been the hopeless dream of astronomers to be able to use telescopes from a point outside the earth's atmosphere.

Edwin Hubble (1889–1953) was an American astronomer who, in 1929, discovered the red shift – the Doppler effect applied to distant galaxies that demonstrated that the more distant the galaxy the more rapidly it was receding. The only possible conclusion from his observations was that the universe was expanding. This finding led to the theory of the 'Big Bang' and to a date – some 15 billion years ago – for the beginning of the universe. When NASA was authorized to place a telescope in orbit around the earth, therefore, and the name 'Galileo' had already been used for the Jupiter spacecraft, there was not too much doubt as to what it should be called.

The Hubble telescope, a massive artificial satellite with an 8 foot wide reflecting telescope was built in the 1980s and achieved two remarkable records – it was at $1.5 billion, rising, as will be seen, to $2 billion – the most expensive scientific instrument ever made, and it provided the most stunning scientific embarrassment of all time. By the end of 1988 the telescope had undergone the most rigorous testing of any device. A few flaws were discovered and the launch was delayed for further testing. In 1989 various major components that had been removed were reinstalled and the fourth major ground system test was performed. In 1990 the satellite was secretly transferred in an Air Force cargo plane from Sunnyvale, San Francisco, to the Kennedy Space Center just two weeks before the devastating earthquake struck the city. Launch was scheduled for March 1990.

On 25 April 1990, after years of delay and repeated testing, the satellite was placed in orbit by the crew of the US space shuttle *Discovery*. During deployment one of its solar panels – necessary for generating power – refused to open properly. Two of the astronauts started to get ready to do a space walk to pull it out by hand, but at the last moment the panel opened up spontaneously. There was another problem. A cable got in the way of an antenna so that its range of movement was restricted. At last, however, the telescope was able to start working.

It was at this point that, to the horror of all concerned, it was discovered that the clarity of the images was little if any better than that of telescopes on Earth. The manufacturer, Perkin-Elmer Corporation, had, apparently, made a mistake in the optical testing. The images were affected by a large amount of spherical aberration and were blurred to about ten times the size had the optics been right. Acutely embarrassed, NASA set up a panel to investigate what had gone wrong.

Incredibly, it found that the all-important 2.4 metre diameter parabolic reflector had been ground to the wrong shape. How was such a thing possible, especially after so much testing? The optics had been assembled and tested in 1981 and it seems that it was in the tests then carried out that the error occurred. A small compensating lens used during the testing had been out of place by 1.3 millimetres. This displacement had, in turn, resulted from an error in interpreting the position of an image produced by an aperture on the end of a measuring rod. The image, it seems, was crucial. The engineers were working to an accuracy of less than a thousandth of a millimetre but all to no avail. Red faces all round.

NASA immediately began to use elaborate computer enhancement methods for the processing and improvement of the images produced by the telescope and the results were so good that embarrassment began to fade. To everyone's relief, the Hubble telescope began to repay the enormous expense and the huge amount of dedicated work that had gone into its design and deployment. There were still technical problems. Power supples were uncertain and the telescope would not readily point itself as intended. At the end of 1993, the most elaborate repair job in NASA's history was mounted when a shuttle visited the telescope, hoisted it inboard with a special crane, replaced part of the lens system to correct the aberration, renewed some of the electronics, replaced the two solar panels and renewed some instrumentation. Some casings, found to be disintegrating, were repaired with Mylar tape the crew had among their supplies.

On 13 January 1994, NASA announced that all had gone well and that the telescope was now working as planned.

> People have been writing about a rocket shot from one country to another, carrying an atomic bomb which would land exactly on a certain target, such as a city. I say that I don't think anyone in the world knows how to do such a thing, and I feel confident that it will not be done for a very long time to come. I think we can leave that out of our thinking.
>
> *Dr Vannevar Bush (1890–1974), computer pioneer, addressing an American Congressional committee in 1945, five years before Russia demonstrated its first intercontinental ballistic missile*

The breeder reactor blunder

Nuclear reactors are essentially simple devices. This is how they work. The atoms of the rare radioactive element, uranium, are constantly undergoing spontaneous breakup (fission). This results in the formation of other lighter elements and the release of uncharged particles called neutrons. Neutrons are released at speed, and if a neutron hits the nucleus of another uranium or plutonium atom it can cause further fission, releasing more neutrons. In this way a chain reaction can occur.

Such a chain reaction produces an immense amount of energy. With each atomic fission a tiny amount of matter is converted into heat energy. Because the energy equivalent of matter is given by the famous equation $E=mc^2$ (energy = mass multiplied by the speed of light multiplied by itself) and c^2 is an incredibly large number, the amount of heat produced is enormous. The energy of the neutrons emerging from a fission is about 80 million times as high as that of atoms in normal matter at room temperature. It is, of course, vital that the chain reaction should be kept under control (see Chernobyl).

Like many other elements, these heavy radioactive elements come in a range of chemically identical forms called isotopes. All the isotopes of an element are chemically identical because they all have the same number of planetary electrons and it is these negatively charged electrons that determine the chemical properties. The number of electrons is the same in any atom of an element because the number of positive particles in the nucleus (protons) is the same. The number of protons determines the number of electrons. This makes atoms electrically neutral. But different isotopes differ from each other in the number of neutrons in the nucleus. Natural uranium is a mixture of two isotopes, by far the greater proportion being the isotope U-238 (with 92 protons and 146 neutrons in its nucleus). Natural uranium can't sustain a chain reaction, for reasons about to be explained. But natural uranium also contains, as one part in 140, another isotope, U-235 (with 92 protons and 143 neutrons in its nucleus).

U-235 is a different kettle of fish. When a neutron hits the nucleus of an atom of U-235 it breaks into two roughly equal fragments and releases either two or three very high-speed neutrons. So if you had a lump of pure U-235 weighing not much more than a kilogram you have an atomic bomb. Such a lump is called a critical mass and, of course, people who make atom bombs ensure that its several sub-divisions are kept well apart until the moment of the explosion.

Natural uranium can't sustain even a slow chain reaction because the preponderance of U-238 in it absorbs fast neutrons too well. U-238 can absorb neutrons to form other isotopes without undergoing fission. When it does this, fissionable atoms are lost. To get a chain reaction going, it is necessary either to slow down the neutrons considerably so that fewer are absorbed by the U-238, or to enrich natural uranium with a higher proportion of U-235. The former method uses a substance called a moderator. In the latter method, using enriched uranium, no moderator is needed.

Nuclear reactors using moderators are called thermal reactors. In these reactors, neutrons are slowed by ensuring that they collide with light moderator atoms such as heavy hydrogen (deuterium), ber-yllium or graphite. Heavy hydrogen is commonly used in the form of. heavy water, deuterium oxide (D_2O). The moderator may be mixed up with the uranium or the uranium fuel may be immersed in heavy water or in another moderator. The fuel rods might fit into a lattice of moderator. This provides a convenient way of controlling the reaction. As the rods are moved in or out of the lattice of moderator, the neutrons are slowed to a different extent. The chain reaction can be stopped if the neutrons are sufficiently absorbed.

The heat produced in the core is used in an entirely conventional manner to generate electricity. Commonly, the heat is allowed to turn water into steam to drive turbines that turn generators. Core heat is transferred to the steam-raising boiler or heat-exchanger by the coolant. The coolant can be a liquid or a gas. Water is frequently used as the coolant and may also be the moderator. Reactors differ mainly in the way the heat is carried from the core to the turbines. There are boiling-water reactors and pressurized-water reactors. In the former, the coolant drives the turbines; in the latter the water has the dual role of coolant and moderator but the primary coolant is not allowed to reach the turbines directly. It raises steam in a secondary circuit and it is this that drives the turbines. In a gas-cooled reactor the coolant is usually carbon dioxide which gets heated to a high temperature. In the advanced gas-cooled reactor the output coolant gas is at 600°C.

Reactors using enriched uranium are called fast reactors because of the higher speed of the neutrons. They don't use a moderator and operate at higher temperatures than the thermal reactors. Because of the higher temperatures, the coolants used are metals that are liquid at reasonably low temperatures. Liquid sodium is commonly used.

So far, so good.

Natural uranium is in short supply and is, of course, expensive. Enriched uranium is even more expensive because the separation of the U-235 from U-238 is very difficult. This is because the two isotopes are chemically identical and differ in their atomic weights by only a tiny percentage. This was the problem facing the British and American scientists when they were trying to make an atom bomb in the 1940s. One of the methods used was to make the natural uranium into the compound uranium hexafluoride, which was then heated to a gaseous form and allowed to diffuse through filters. The slightly lighter U-235 passed through a little more easily than the U-238, thereby very slightly enriching the mixture on the other side. Several thousand filter stages occupying a site covering many acres were needed to produce the required purity of U-235. Two other methods were also tried, one of them a nuclear reactor that produced a brand new man-made element that would fission. In 1944 the Manhattan Project was spending money at a rate of more than $1 billion per year.

In the light of all this difficulty and expense the possibility of a type of nuclear reactor that actually produces more usable fuel than it consumes is an attractive proposition. Such a type is the breeder reactor. Breeder reactors are high-speed reactors in which the neutron speeds are not slowed down to thermal velocities as in thermal reactors. Their fuels are either U-238 or thorium-232. The most effective type of breeder reactor seems to be the liquid-metal fast breeder reactor, which converts U-238 into a new, man-made element called plutonium-239. Plutonium is readily fissionable and will easily set up chain reactions. It was plutonium that was finally used to make the bombs that destroyed Hiroshima and Nagasaki. Here is how plutonium is formed.

An atom of U-238 with its 92 protons and a total of 238 protons and neutrons is struck by a high-speed neutron. This neutron is captured to make another uranium isotope U-239. It is still uranium because the number of protons is unchanged at 92. U-239 is, however, very unstable and one of its neutrons shoots off a high-speed electron called a beta particle. You can quite legitimately think of a neutron as being made of an electron (negative) stuck firmly to a proton (positive). This makes it neutral. Remember that we are considering a particle in the *nucleus* of the atom, not one of the planetary electrons. An electron coming from a nucleus is called a beta particle. If a neutron gives off an electron, it will, of course, lose a negative charge and turn into a proton. So the atom is no longer uranium. It now has 93 protons and immediately acquires an extra planetary electron to become a chemically different element. It is, in fact called neptunium-239. This, too, is a very short-lived element and it quickly gives off a beta particle. The result is another transformation to yet another element, plutonium, with 94 protons and a total of 239 nuclear protons and neutrons.

All this can be made to happen in a fast breeder reactor so that the final result is a fuel of 'spent' U-238 with which is mixed up a large quantity of plutonium. The energy tied up in the plutonium is about 70 times the energy produced in the reaction that forms it. Removing this plutonium from the spent fuel is called reprocessing, and this is what has been going on at Sellafield since 1994.

For all its value as a nuclear fuel, plutonium is probably the nastiest substance known to humanity. The half-life of plutonium-239 is 24,360 years. That means that a lump of plutonium will take 24,360 years to fission away spontaneously to half its original bulk. A kilogram made this year will be reduced to half a kilogram by the year 26,360. This is a nice inheritance for our descendants (if any). As J. T. Edsall wrote in the book *Environmental Conservation*: 'We have entered into a Faustian bargain whereby we are given an unlimited energy source in return for a pledge of eternal vigilance.'

Plutonium is well named. Pluto was the Greek god of the underworld – the hell to which the dead were carried. Plutonium is a silvery metal that develops a yellowish tarnish on exposure to air. It is constantly warm because of its radiation. It is also one of the most toxic and hazardous substances known. It does not occur in nature, but people are making it all the time, and it is for practical purposes permanent. It can be destroyed only by using it as an enrichment for U-238 in nuclear reactors or by making 'dirty' atom bombs of it and setting them off. To do this would be to disperse a lot of plutonium around the world. There is no other known way of disposing of it, except, perhaps, by shooting it off into outer space in a rocket. Plutonium is a bone-seeking poison similar to radium but several times as toxic. When plutonium is burned, as in a fast breeder nuclear reactor accident, the compound plutonium dioxide is formed. This would be released as an aerosol of very fine particles – some 10,000 million per gram of the metal. Each one of these particles is estimated to carry a 1 per cent risk of causing lung cancer in anyone who inhaled it.

In the early 1970s Britain, America, France, West Germany and the USSR all built prototype fast breeder nuclear reactors that were generating plutonium. Although none of these countries has exploited fast breeder reactors commercially, <u>the world's stockpile of plutonium</u> is increasing steadily. <u>Today (1998) it is about 1700 tonnes.</u>

<u>The critical mass of plutonium</u> – the amount that will explode as an atomic bomb when brought together – <u>is less than a kilogram.</u> The implications of this for international terrorism hardly bears thinking about. A bomb that could destroy a city could be made to fit in a small car. Plutonium has gone missing from government nuclear weapons plants. In May 1994 German police found 6 grams of highly purified plutonium-239 during a raid on a businessman's home. In July the German government confirmed that this plutonium probably came from Russia. Helmut Kohl complained to Boris Yeltsin and a Russian commission was set up to investigate the smuggling of fissionable material out of Russia.

Cars will cost as little as $200. People will have two-month vacations. They will care little for possessions. The happiest people live in one-factory villages.

Predictions for 1960 by General Motors in a 'Futurama' exhibit at the 1939–40 New York World's Fair

---

Science and Religion

> The earth was created in the year 4004 BC
>
> *Bishop James Ussher (1581–1656)*

Don't tangle with the Church

Giordano Bruno (1548–1600) was an Italian astronomer, mathematician and scientist whose far-seeing scientific imagination was two centuries ahead of his time. He proposed that the universe was of infinite size; that the earth was one of numerous possible worlds; that the sun did not move round the earth; and that the earth was not the centre of the universe.

Bruno, although amazingly indiscreet, was one of the great heroes of science and it is a pity that he is so little known. He was the son of a professional soldier and, in 1565, entered a Dominican convent in Naples. But he soon found that many of the theological matters he was expected to study simply didn't make sense. His main doubts were on the topics of transubstantiation and the immaculate conception. Transubstantiation is the Roman Catholic doctrine that the bread and wine, when consecrated in the Eucharist, actually and literally changes into the substance of the body and blood of Christ, regardless of the evidence of the senses. The immaculate conception is the Roman Catholic belief that the Virgin Mary was, from the moment of her own conception, free from any taint of original sin – that is, free from any of the theologically negative aspects of sexual experience.

Since he was not able to keep his unorthodox ideas to himself, Bruno was soon suspected of heresy. In spite of this, however, after seven years' study he was ordained as a priest. Even then he continued to question religious dogma, and eventually he was arrested and charged with heresy. Bruno was well aware of the danger he was in, so before he could be tried by the Naples Inquisition he escaped to Rome. There, almost at once, he was unjustly accused of murder. Another heresy trial was set up and again he had to make a run for it.

Having decided to abandon the Roman Catholic Church, Bruno enlisted in the Calvinist persuasion. But, perhaps unsurprisingly, he again found himself at loggerheads with the dogmatists and was arrested, tried for heresy and excommunicated. This grave sentence meant that when he died he would go to Hell, so Bruno recanted and was restored to the fold. He was then allowed to move to France, where he got a job as a philosophy lecturer in Toulouse, and then in Paris, where he soon gained the favour of the king. In 1582 he published three philosophical works concerned with acquiring knowledge of reality, and a stage comedy that was really a protest against the corruption of the times. The following year he moved to Oxford, where, thinking it might be safe to do so, he lectured on the Copernican theory that the earth moved round the sun. He denied that the earth was flat and was supported on pillars. He denied that the sky was a firmament fixed above the earth and was the floor of Heaven. It was, he claimed, an infinity filled with self-illuminated worlds many of which were inhabited. There was nothing above us but space and stars.

Bruno was arguing against the teachings of Aristotle, and, since Aristotle had been approved by the Church, his teachings almost had the force of holy Scripture. Bruno's optimism was unfounded; a serious quarrel with Oxford scholars followed and Bruno had to retire to London as the guest of the French ambassador. There he was received at the Court of Queen Elizabeth I and met many of the important men of the day. Encouraged, Bruno wrote a set of six dialogues laying down his scientific and cosmological belief. These dialogues contain much that has since been proved to be true. Not content with this Bruno included in these writings the outrageous suggestion that the Bible was not to be considered as a textbook of science. He also criticized contemporary superstition and the belief in religious salvation by faith alone.

In October 1585 Bruno returned to Paris, where he found that, as a result of political change, he no longer enjoyed royal protection. Unwisely, he again publicly attacked Aristotle's doctrines and expounded his own remarkably enlightened idea of religion. This involved mutual respect and understanding between all religions and their peaceful coexistence. He was again tried and again excommunicated. By now Bruno had piled up a fatal amount of evidence against himself. He also had a remarkable talent for making enemies by his unremitting tendency to declaim against the hypocrisy and insincerity of his religious persecutors. He repeatedly insisted that he was not fighting against men's beliefs but against their pretended beliefs. He simply could not credit that intelligent people could genuinely hold to the propositions they were professing.

Had he remained in France, all might have been well. But in August 1591 he rashly accepted an invitation to return to Italy. Hoping to be appointed to the vacant chair of mathematics at the University of Padua, he went there and began a course of lectures. His application, however, failed and the post was offered to Galileo. So

Bruno went to Venice as the guest of one Mocenigo who hoped to enjoy great advantages from Bruno's private instruction on the art of memory. This hope was unfulfilled and in May 1592 the disappointed Mocenigo denounced Bruno to the Inquisition for his heretical theories. Bruno was arrested and tried for heresy. He seemed to be defending himself quite successfully when the Inquisition in Rome demanded his extradition.

Bruno was taken to Rome, locked up in the jail of the Holy Office, and for seven years was without books, paper or friends as he stood trial. His defence was that his interests were scientific and philosophical and not theological. But the inquisitors demanded that he must unconditionally retract all his scientific theories. Bruno then tried to show that these scientific beliefs were not incompatible with a Christian belief in God and creation. He believed in an intellect which animated the universe and that the visible world was a manifestation of this great intellect. This all-pervading intellect was God. These views cut no ice with the Inquisition and Bruno was repeatedly pressed to deny and retract his science. He refused.

Bruno was convicted of heresy, specifically by teaching the plurality of worlds – 'a doctrine repugnant to the tenor of Scripture and contrary to revealed religion and especially to the plan of salvation.' Pope Clement VIII therefore ordered that he be sentenced as an 'impenitent and pertinacious heretic'. On 8 February 1600, the sentence was read to him. He was to be delivered to the secular authorities to be punished 'as mercifully as possible, and without the shedding of his blood'. This was the comfortable euphemism for burning at the stake.

When Bruno was sentenced he said to his judges: 'I suspect that my fear in receiving this judgment may be less than yours in passing it'. To stop him from talking, his tongue was fixed in a gag and he was taken to the Campo de' Fiori where he was tied to a post and burned alive.

Let him be anathema who unblushingly affirms that, besides matter, nothing else exists. Let him be anathema who shall say that no miracles can be wrought, or that they can never be known with certainty.

Pope Pius IX (1792–1878)

Galileo and the Inquisition

The Italian scientist, writer, inventor and professor of mathematics Galileo Galilei (1564–1642) was a man of enormous erudition and clarity of mind. He was an artist, a poet and a writer with a brilliant literary style. He wrote essays on Dante, Ariosto and Tasso. He loved music and was a virtuoso player on the lute. He also had a wonderful sense of humour which endeared him to his students. As a teenager, he went to the University of Pisa to study medicine. There he noticed that a swinging pendulum always took the same time for each excursion and he used this principle to invent a machine that could count heart beats and show the rate on a scale. Soon after starting, he read a book by an older contemporary Giordano Bruno (1548–1600) (*see* Don't tangle with the Church) which explained the Copernican theory of the universe. Galileo was fascinated by this, gave up the idea of becoming a doctor, and resolved to be a scientist.

From the writings of Nicolas Copernicus (1473–1543), who died before Galileo was born but is now recognized as the father of modern astronomy, he learned some wonderful things. The Copernican theory was nothing less than a complete revolution in human thought. At the time, the great authority on all scientific matters was the Greek philosopher and scientist Aristotle (384–322 BC). In those days, scientists were not obsessed with the idea that newest was best and were not in the least disturbed that their latest authority had lived nearly 2000 years before. Aristotle had become a kind of oracle who could not be mistaken. According to Aristotle – and his views seemed to almost everyone to be the plainest common sense – the earth was the centre of the universe and the sun moved round it, thereby causing day and night.

But Copernicus had good reason to think differently. On the assumption that the earth is the centre of the planetary system, the motion of the other planets appeared very odd indeed. It was clearly observable that they did not move in circles round the earth as the sun appeared to do. Instead, they moved in a series of loop-the-loops, each loop taking exactly one year. The motion was like that of a point on the edge of a wheel as the wheel rolls along a surface. In the case of Jupiter, there were a little over 11 loops in the entire cycle which took 4332 days. In the case of Saturn, the whole cycle took 28 years. This extraordinary motion of the planets was well known to the ancients, as it was a simple matter to plot the apparent position of a planet on the sky. No one seemed to think it odd that a planet should travel forward, then loop and travel backward, then forward again every year.

No one but Copernicus. This Polish astronomer, medical man, canon lawyer and military governor was a believer in simplicity. He also had a lot of common sense and a remarkably open mind. So he set himself to see whether there was a better explanation for the seeming movement of the planets. In the end, he found it, but it involved a remarkable paradigm shift no less than the proposition that the earth was not the centre of the universe. Copernicus showed, by unequivocal geometric diagrams, that if the sun was taken to be the centre of the system of planets and if it was accepted that the earth and all the planets moved in orbits round the sun, and the earth was rotating on its own axis, all the complexities would be swept away.

Copernicus put all this down in a book called *De revolutionibus orbitum coelestium*. Ironically, Copernicus dedicated his book to the Pope and persuaded a cardinal to pay for the printing of it. So this book, which was to cause so much trouble to the church, was actually sponsored and published by the church. Perhaps fortunately for Copernicus as later events were to suggest, the book came out shortly before his death. It was left to Galileo to bear the brunt of the theological upset this book was to cause.

Galileo made a good start in stirring up trouble. While still a young man, he pointed out that Aristotle was wrong in saying that a heavier body fell faster than a light body. Not content with this, he set about proving it by experiment. This extraordinary notion of trying to disprove the teachings of Aristotle, which had come to be believed almost as religious dogma, attracted unfavourable attention in Pisa and there were mutterings against him. Fortunately, his reputation and status were such – he had been appointed professor of mathematics at Pisa – that he was offered a chair in Padua where his lectures attracted students from all over Europe.

In 1609, having heard that it was possible, by means of two lenses, to see distant objects as if they were much closer, Galileo decided to experiment. After many trials with lenses ground with his own hands, he developed a telescope with a convex object glass and a concave eyepiece – a system known to this day as the Galilean telescope – that could magnify 30 times. This instrument had the advantage over the earlier form with two convex lenses that it did not invert the image. The telescope caused a sensation and everyone of importance in

Padua came to see through it. Galileo, always quick to take a hint, made one for the Senate who immediately doubled his salary to 1000 florins and agreed that it would be paid for life.

When he looked at the moon with this instrument, he was astonished to see that it had mountains and valleys. Using an ingenious method, he measured the height of the mountains on the moon. He then examined the constellation known as the Pleiades and, instead of seven stars as was believed, he found that there were more than forty. Looking at areas of the sky in which no stars were visible with the naked eye, he discovered that many stars could be seen with his telescope. To his astonishment he was able to make out that four tiny bodies were actually revolving around Jupiter. He showed that there were spots on the sun's surface and that it was rotating.

Most important of all, a careful inspection of the planet Venus showed that it had phases, exactly like those of the moon. At different times it showed as a crescent, then half-sphere, then gibbous, then full. Galileo was quick to recognize that this was clear proof that the light from the planet was reflected rather than self-produced and that it was rotating around the sun. The Copernican theory must be right. Significantly, the principal professor of philosophy at Padua, although repeatedly invited to verify these findings for himself, categorically refused to do so, preferring merely to repeat his reasons for disbelieving them. This is an excellent example of how dogmatic belief can take precedence over the desire for truth.

The news of Galileo's findings reached Rome and caused consternation among the theologians. The stars had been made by God to provide illumination for mankind. This was clearly stated in the Bible. How could there be stars that were invisible to the naked eye? Man, on earth, was at the centre of the universe. The sun moved round the earth, as was obvious to anyone. Galileo's telescope must be a fraud. He was an impostor, a blasphemer, a heretic and an atheist. Galileo was summoned to Rome.

At first he was well received. He had a long and friendly audience with the Pope and they parted on excellent terms. He showed his telescopic observations to many of the churchmen and they seemed to agree with his findings. But the dogmatic theologians in the College of Cardinals were now in a quandary. The Bible was literally true and this was a fact they could not set aside. To accept what Galileo was teaching was unthinkable. So, finally, dogma prevailed over clear evidence and Copernicus's book was declared anathema and was placed on the *Index librorum prohibitorum* – the list of forbidden books. Galileo was formally commanded never to teach that the earth moved.

Horrified and alarmed, he wrote to Father Castelli to suggest that the Bible had never been intended to be a scientific treatise but was to be considered a moral guide. This was a serious mistake. It was no business of a mere layman to comment on the nature of divine scripture. Galileo was summoned before the Inquisition and was accused of the grave offence of declaring that the earth moved round the sun – an assertion directly contrary to the bible. On pain of imprisonment, he was ordered to renounce his heresy and to refrain from teaching or even defending the principle that the earth moved round the sun. Galileo, knowing that he was risking torture and execution, promised to obey. He was well aware of his danger. His scientific hero Giordano Bruno, whose book had led him into science and who had championed the Copernican theory, had been arrested by the Inquisition in 1592 and, after a seven-year imprisonment and trial, had been burned at the stake (*see* Don't tangle with the Church).

For 16 years Galileo was silent on the Copernican question, knowing that it was not only the Church that was against him.

Most of his scientific colleagues, especially those in the universities, were also ready to condemn him. The ideas he had put forward were deeply contrary to the received scientific wisdom of the time. Aristotle was right, he was wrong, and Copernicus was a fool. The Church in its wisdom agreed with Aristotle, so he must be right. So Galileo devoted his time to less sensitive scientific pursuits. He invented a thermometer and a calculating compass, made major advances in hydrostatics, and developed the compound microscope.

Galileo was an immensely popular lecturer and a brilliant arguer. His method with hostile disputants was to lead them on, helping them to explain their position as clearly as possible and often bringing out points in their favour that they had not seen. Then, when it seemed that he had almost defeated himself, he would proceed to undermine and then utterly demolish their arguments. These tactics delighted his students who spread his teachings widely. Among many other scientific achievements, he discovered the laws of motion and laid down the foundations of the science of mechanics.

In 1623 the old Pope died and his successor, Urban VIII, showed great favour to Galileo, especially after he had gone to Rome to congratulate the new Pope on his elevation. Encouraged, Galileo set about the writing of a book *Dialogues on the Two Chief World Systems*. This entertaining work was an account of conversations between a Copernican scientist, Salviati, an impartial and witty scholar Sagredo, and an absurd, ponderous and hopelessly stupid Aristotelian, Simplicio, who comes out with all the stock arguments. The book provided clear and unassailable proof that the Copernican theory was correct and left no doubt as to the author's beliefs. Simplicio's arguments are systematically ridiculed and refuted by the other two until he is reduced to baffled rage and impotence. In the preface Galileo states that his purpose was to show that the edict condemning the Copernican theory had not, as many believed, been made in ignorance of the facts, but that the churchmen clearly understood the arguments for them. For two years he was unable to get the book printed but, eventually, because of the carelessness – or idleness – of the ecclesiastical censor the book was passed. The unfortunate censor was later dismissed.

The book was published in 1632 and at once became very popular. The attempts of the Church to suppress it only added to its success. The Cardinals were furious and represented to the Pope that Galileo intended it to be understood that he, himself, the Holy Father, was Simplicio. Nothing could have been further from Galileo's intention, for he was a close personal friend of the Pope and grateful to him for his former support. But Urban was persuaded and, within months of its publication, the sale of the book was prohibited and its text submitted for examination by a special commission. Inevitably, this committee reported against Galileo.

Galileo, then nearly 70 and frail and infirm, was again summoned to Rome to appear before the Inquisition. He was told that he had brought on himself the penalties of heresy. Several times he was examined and each time he risked rigorous examination – the euphemism for torture – unless he formally recanted. Galileo thought of Giordano Bruno. His friends urgently advised him to comply. The threat undermined his health. This went on for four months and then he was told that he would be required for rigorous examination the next day. He went and remained for three days behind closed doors. There were no outside witnesses of what happened and Galileo was sworn to secrecy. The records of the Inquisition are never revealed.

We do know, however, that people in his terrible situation were subjected to five stages. First, they were officially threatened in court. Second, they were taken to the door of the torture chamber and the threat was repeated. Third, they were taken inside and shown the instruments of torture, Fourth, they were undressed and bound upon the rack. Fifth, they were physically tortured. There is no public knowledge of how far these people went with Galileo. We do know, however, that, afterwards, he was found to have developed a hernia – a common, and well-known, consequence of physical torture.

Galileo recanted. He was dressed in the clothes of a penitent and brought before the Cardinals and prelates for judgement. He was made to kneel and to put his hand on a Bible. He was required to abjure and curse the doctrine that the earth moved round the sun. This he did, stating: 'I Galileo Galilei, aged seventy years . . . having been judged grievously suspected of heresy; that is to say that I held and believed that the sun is the centre of the universe and is immovable, and that the earth is not the centre and is movable . . . I abjure, curse and detest the said errors and heresies . . . and I swear that I will never more in future say or assert anything verbally, or in writing, which may give rise to a similar suspicion of me; but if I shall know any heretic, or anyone suspected of heresy, that I shall denounce him to this Holy Office. But if it shall happen that I violate any of my said promises, I subject myself to all the pains and punishments which have been decreed and promulgated by the sacred canons against delinquents of this description.'

He was then sentenced to imprisonment for the rest of his life.

Legend has it that Galileo, on rising, muttered 'But the earth still moves.' This is almost certainly apocryphal. Such a remark, had there been anyone to hear it, would have been fatal, and Galileo was, by then, a broken and friendless old man, utterly incapable, from the suffering he had undergone, of offering any further resistance or of taking any further chances.

His persecutors returned home, satisfied that they had done their sacred duty. Galileo was forbidden to write or publish anything. He was denied the right to receive friends or to teach. Many of his manuscripts were burned.

Wherein lies the blunder in this terrible story of the destruction of one of the great men of all time? No disinterested person could deny the clear evidence in favour of the Copernican theory. But the churchmen could not be disinterested. Their ideas and principles were central to their whole lives. Much the same applied to the Aristotelian scientists and philosophers. Both groups had a fixed, complete and unalterable system of truth, and both, for different reasons, found any criticisms of their systems intolerable. They were ready to persecute anyone who seriously questioned their systems, especially anyone who brought clear evidence to show that they were wrong. Significant or not, there was a rather long interval before the Church formally admitted that they had been wrong in forcing Galileo to deny the evidence of his – and everyone else's – senses. Just over 350 years to be precise. This conciliatory event occurred in 1992.

So the blunder has nothing to do with stupidity; it consists in the refusal to accept the possibility that a widely accepted and agreed system – a system that seems to account comfortably for observed phenomena – could be wrong. It is tempting to think that, today, we have progressed beyond the stage of dogmatic assertion, that we are completely open-minded and ready to adjust to all new scientific ideas. But this is not so. The whole history of science, right up to the present, is a story of refusal to accept fundamental new ideas; of determined adherence to the *status quo*; of the invention of acceptable explanations, however ridiculous, for uncomfortable facts; of older people of scientific eminence dying in confirmed possession of their life-long beliefs; and of painful readjustment of younger people to new concepts.

> Air is necessary for violent motion. Violent motion exists between the moon and the earth. Therefore the space between the moon and the earth is full of air.
>
> *Aristotle (384–322 BC)*

Heart and soul

There is a strange anomaly in the use of the word 'heart', in the sense of being the seat of one's innermost thoughts and feelings. In this sense, the heart is treated, not as an anatomical structure, but as a mystical entity corresponding to the soul or the spirit. In a sermon preached in 1635 Robert Sanderson (1587–1663), Bishop of Lincoln, said: 'The heart is very often in Scripture taken more largely, so as to comprehend the whole soul, in all its faculties, as well the apprehensive as the appetitive; and consequently taketh in the thoughts, as well as the desires, of the soul.'

This usage is not, however, simply figurative, for everyone who employs it is aware of the heart as a physical entity. 'You are breaking my heart,' a mother might say to her wayward son, and in so saying is employing an image, however metaphorical, that implies something capable of being physically broken. The heart is also often taken to be the seat of the emotions, as distinguished from the intellectual nature – which is located in the head. 'Her head told her that the man was contemptible; but in her heart she knew that she loved him.' People talking in this way are, we must presume, well aware that the brain is the seat not only of the intellect but also of the emotions. They may also be well aware that the heart is simply a muscular bag that pumps blood round the body.

So it is clear that there is a strong human need for an imaginary organ that has the characteristics so commonly attributed to it. This organ has nothing to do with the brain, and it is not necessarily identified with the soul. Theologians, and others, commonly talk of 'heart and soul' – a usage that would be unnecessary if the two were one. The heart is the seat of courage, mercy and pity. 'Have a heart' we say. It is the seat of kindliness, as the term 'cordiality' implies. It is also the centre of temperament and character, to say nothing of purpose, intention, inclination, will and desire. It is, of course, the seat of love and affection.

This extraordinary duality of significance must have arisen at a very primitive stage in human evolution, long before the anatomical and physiological features of the heart were known. The German theologists Karl Rahner and Herbert Vorgrimler have suggested that the physical heart was, from the earliest times, the symbol of the 'real' heart. The latter, they say, is 'the dynamic principle which drives man to seek that ultimate and ultimately unattainable understanding of himself which can only be found in his own heart.' If you can understand that, you are a better man than me, Gunga Din.

Philosophy of Science

> 'A body in motion can maintain this motion only if it remains in contact with a mover.
>
> *Aristotle (384–322 BC)*

How to detect pseudoscience

There is an excellent book called *Conjectures and Refutations* (1963) by the Austrian-born British philosopher Sir Karl Raimund Popper (1902–94). The chapter from which the book takes its title is concerned with how to distinguish real science from pseudoscience. The chapter is based on a lecture on the philosophy of science that Popper gave at Peterhouse, Cambridge, in 1953 under the auspices of the British Council. The need to distinguish pseudoscience from science is even more important today than it was then.

Pseudoscience is alive and well. Popular journalism, the astonishing achievements of science of the past 50 years, the desire to seem to share in this status, and perhaps even the profit motive have encouraged many pseudoscientists to claim to be scientific or to be engaging in scientific activity. It is very important for us, exposed as we are, to an immense volume of claims of the validity of pseudosciences such as astrology, various kinds of supplementary medicine, psychoanalysis, economics, clairvoyance, metaphysics, and many others, to know when we are being conned or, at least, misled.

Popper, who had been working on this problem since 1919, was, at first, surprised to find that one of the features of those entities he believed to be pseudosciences was the remarkable power they had to explain everything. In his paper he looks particularly at Karl Marx's theory of history, Freudian psychoanalysis and Adler's 'individual psychology'. These theories could explain everything, and did. Anyone professionally engaged with them could be relied on to produce a constant stream of observations that 'verified' the particular theory.

This was, and is, especially true of the psychoanalysts. Popper was an acquaintance of Alfred Adler and discussed with him the case of a child, whom Adler had not seen, but whose problem he had no difficulty in explaining in terms of an 'inferiority complex'. Popper, who was not in the least convinced by the argument, was shocked and asked how Adler could be so sure he was right. 'I know I am right,' said Adler, 'because of my thousand-fold experience.'

It then occurred to Popper that none of Adler's previous experiences might well have been any sounder than this one was, and that each in turn had been interpreted in the light of 'previous experience' while being used to increase the mountain of 'evidence'. It was clear to Popper that all that Adler had proved was that this case could be interpreted or described along the lines of Adler's theory of the inferiority complex. Since, however, *any* case, whether manifesting indications of superiority or feelings of inferiority, could be interpreted in this way, its value as evidence of truth was virtually non-existent. Soon it became apparent to Popper that, far from being a strength to assert that a theory could explain everything, it was actually a fatal weakness.

Commenting on this, Peter Medawar, the Nobel Prizewinning scientist and highly entertaining writer, cites a number of hilarious passages taken from the proceedings of the 23rd International Psycho-analytic Congress, held in Stockholm in 1963. The passages quoted by Medawar were from the authors' own summaries of their contributions to the Congress. One contributor explained anti-Semitism thus: 'The Oedipus complex is acted out and experienced by the anti-Semite as a narcissistic injury, and he projects this injury upon the Jew who is made to play the role of the father. His choice of the Jew is determined by the fact that the Jew is in the unique position of representing at the same time the all-powerful father and the castrated father.'

Another contributor explained why a man suffering from ulcera-tive colitis dreamed of snakes: 'The snake represents the powerful and dangerous (strangling), poisonous (impregnating) penis of his father and his own (in its anal-sadistic aspects). At the same time, it represents the destructive, devouring vagina. The snake also repre-sents the patient himself in both aspects as the male and female and serves as a substitute for people of both sexes. On the oral and anal levels the snake represents the patient as a digesting (pregnant) gut with a devouring mouth and expelling anus.'

Yet another account of a manic-depressive patient, who had been in analysis for six years, reads: 'The delusion of having black and frighten-ing eyes took the centre of the analytic stage following the resolution of some of the patient's oral-sadistic conflicts. It proved to be a symptom of voyeuristic tendencies in a split-off masculine infantile part of the self and yielded slowly to reintegration of this part, passing through phases of staring, looking at and admiring the beauty of women.'

Examples of this kind can be multiplied indefinitely. There are endless tomes devoted to this kind of self-indulgent nonsense. But the importance of them in this context is that, although these analysts are dealing with problems of the greatest medical complexity and difficulty, none of them shows the smallest indication of doubt or hesitancy in their assertions. Once the jargon has been mastered and a few 'principles' grasped, there are no further problems in explaining anything. One of the triumphs of psychoanalysis is to be able to explain why it is that some people don't believe in it. It is, *par excellence*, the *argumentum ad hominem*. People who don't believe in psychoanalysis are mentally sick. They may have one of a range of problems, but it is usually an unresolved Oedipus complex.

This is, perhaps surprisingly, a circular argument. As Medawar points out, a system that can explain everything, including why some people don't believe in it has, paradoxically, deprived itself of the power to explain why some people do. You can't logically subject a system like psychoanalysis to a critical scrutiny by using psycho-analytic techniques to make the scrutiny.

Popper goes on to compare this kind of thing with a real scientific theory and makes the point that while, in their assertions, the psycho-analysts are taking no risks, the scientist who makes a prediction takes a big risk of being proved wrong. Whatever the outcome of any psycho-analytic activity – whether the patient gets better or worse – the psychoanalysts always have an explanation. But if a scientist makes a bold assertion about a matter that can be tested by experiment, he or she has to stand by the outcome of the experiment and be prepared to give up any theories that are falsified by the results.

Considering astrology, Popper found that astrologers could pro-duce an immense amount of confirmatory evidence linking birth date and time to subsequent character and personality development of the kind predicted by astrological theory. What they failed to do, however, was to pay any attention to the equal mass of evidence that showed no correspondence whatsoever between birth date and character or personality. Considered honestly, the totality of the evidence would have proved nothing. More significantly, perhaps, the astrologers were shown to be remarkably skilful in producing predictions and interpretations so vague that they could be used to explain away any evidence that went against their theories.

So far as the Marxian theory of history is concerned, Popper found that some of the early assertions were, in fact, susceptible to proper testing and were, incidentally, shown to be false. Later, however, Marxist theoreticians engaged in a reinterpretation of both the theory and the evidence so as to make them agree. In this way they produced a system that was self-proving, whatever happened.

So what is the unifying principle behind all this that allows us to detect pseudoscience? It is simply this: the difference between a science and pseudoscience is that scientific statements can be proved wrong and pseudoscientific statements cannot. By this criterion you will find that a surprising number of seemingly scientific assertions – perhaps even many in which you devoutly believe – are complete nonsense. Rather surprisingly this is not to assert that all pseudo-scientific claims are untrue. Some of them may be true, but you can never know this, so they are not entitled to claim the cast-iron assurance and reliance that you can have, and place, in scientific facts.

> It would be difficult to cite any proposition less obnoxious to science than that advanced by Hahnemann: to wit, that drugs which in large doses produce certain symptoms, counteract them in very small doses.
>
> *George Bernard Shaw (1856–1950)*

Medicine off the rails

No book on scientific blunder would be complete without at least a brief reference to one of the of major blunders in the history of

<u>medicine</u>. Ironically, this particular blunder arose for the best of all reasons – the desire to do something about the appalling state of medicine at the time.

In 1780 medical treatment was primitive in the extreme. There were only half a dozen drugs that were of any real use. Surgery was at the level of butchery and most patients who underwent major surgery died after great suffering. The commonest therapeutic measure, bleeding, was totally useless and was, in fact, responsible for the deaths of enormous numbers of people who desperately needed all the, often anaemic, blood they had.

Just about the only thing you could say for the medical treatment of the day was that it was, at least, based on some measure of common sense. If something was wrong, the reasonable thing to do was to find a way of opposing it. To most people that would seem no more than elementary logic. The fact that, at the time, hardly anything was known about the causes of disease, is neither here not there. But there was one man who did not see things that way. Many people call him a genius. His ideas certainly caused a strong impression – so strong, indeed that many people believe in them to this very day. That man was <u>Samuel Hahnemann (1755–1843).</u>

As a young doctor, Hahnemann was disgusted with the way doctors were behaving. In particular, blood-letting seemed singularly pointless. Hahnemann may have recognized that people who got better after bleeding did so in spite of it rather than because of it. He was certainly bright enough to come to this conclusion. Although of fairly humble origins – his father was a Meissen porcelain painter – he had been a brilliant student and a master of French, Latin, Greek and Hebrew. He had also taken a great interest in science, especially in chemistry and botany. As a medical student, he wrote a dissertation in Latin on the anatomy of the hand. Eventually, however, his disgust with medicine led Hahnemann to give up clinical practice and to devote himself to translation of medical and other texts, general writing and scientific study.

In 1790 Hahnemann was translating *Materia medica* – a book of drug action – and was struck by the different explanations given of the action of Peruvian bark in fever. Peruvian bark is cinchona and we now know that it contains quinine. This is a drug that attacks the malarial parasite and that has saved countless lives by relieving or suppressing malarial attacks. At the time, doctors had no idea about the cause of malaria. It was simply called 'fever' and, apart from the known fact that the bouts of fever occurred at regular intervals, was not distinguished from other causes of raised temperature. No one could possibly have known how Peruvian bark worked.

Hahnemann decided to try Peruvian bark on himself and took a large dose. According to all the available accounts of the affair and his own recorded documentation, he was struck by the similarity between the effects the drug had on him and the symptoms of intermittent fever. Malaria causes bouts of fever and flushing, with shaking, severe headache, muscular aches and pains, profuse sweating, weakness and debility. Quinine in large doses causes headache, nausea, ringing in the ears, deafness, blurred vision, double vision, night blindness and even permanent loss of vision. Fever and flushing of the skin are not features of quinine intoxication except in rare cases of people who are hypersensitive to the drug. Perhaps this was the case with Hahnemann.

Hahnemann was so impressed by these alleged similarities that he decided to try other drugs and to see how their effects compared with the symptoms of the diseases they were used to treat. This was an interesting line of enquiry, and such as would be applauded by any scientist. Up to this point Hahnemann was behaving rationally. But now, his desire to found a new basis for medical treatment must have begun to affect his detachment and judgement. People with something they wish to prove will commonly develop a kind of selective awareness. This takes the form of blindness to any evidence that opposes the pet theory and enthusiastic adoption, recording of, and reference to all evidence that seems to support it. It is highly effective in moulding strongly held opinion on matters for which there are no real grounds for belief.

There is no way that Hahnemann could have found enough real evidence to establish the doctrine of '*simila similibus curentur*' (like cures like), which he was shortly to propose. For a start, the enormous majority of the range of drugs that were then in use had no medical value whatsoever. Their adoption had been based on nothing more than the observation that some people who had taken them had recovered from their illness. Since the immune system sees to it that we recover from most illnesses whatever we take, Hahnemann's naive belief in the genuine therapeutic link between certain drugs and the cure of certain diseases was without foundation.

There was, moreover, only a handful of drugs that had any useful effect whatsoever on the body. There was opium, which is a pure pain-killer with no specific curative effect on any disease; there was quinine, which is useless in any feverish illness other than malaria; there was mercury, a very weak and unsatisfactory remedy for syphilis; there were cascara sagrada and castor oil for constipation; there was arsenic, which, in the form then available, had no therapeutic justification in any disease; and there was belladonna (atropine), a highly poisonous remedy, effective in relieving colic but, at the time, used mainly to widen ladies' pupils and make them look more beautiful (hence the name). It is doubtful if Hahnemann knew about digitalis from foxglove which had been introduced by the English physician William Withering in 1785. That was about the extent of the useful drug armamentarium at the time, so Hahnemann had nothing like the number necessary to prove his case.

Again, those drugs with useful effects were not simply used on one single disease per remedy. Opium, for instance, was considered a 'specific' for a wide range of painful conditions. Medical knowledge at the time was not always clear about the distinction between diseases and symptoms and much of the medical treatment was directed at the removal of symptoms. In fairness to Hahnemann, it must be said that he was aware of this, and was seriously looking for more than simply palliative treatment. He wanted to find genuine remedies that, as he put it, 'aim at rooting out the evil by specific means'.

Unfortunately, these few observations seemed to Hahnemann to provide a clue to what was to become a completely new system of medicine. His idea was that drugs that produce symptoms similar to those of a disease are the very thing needed to cure the disease. As he said in his principal book, *The Organon of the Rational Art of Healing*, 'A disease can only be destroyed and cured by a remedy which has a tendency to produce a similar disease, for the effects of drugs are in themselves no other than artificial diseases.' To describe this process Hahnemann invented a word 'homoeopathy' derived from the Greek *homos* meaning 'same' and *pathos* meaning 'disease'.

So far, so good. This was a reasonable proposition – untrue, but also untested and worthy of trial. At this point, Hahnemann ran into a problem. If the drugs he wanted to use simply caused the diseases he wanted to cure, of what possible benefit could they be? This was a poser. There was only one way out of the difficulty. Since he must not do anyone any harm, he must give these drugs in a very small dose. When he did this he found, perhaps unsurprisingly, that: 'the smaller the dose of the homoeopathic remedy, the less will be the aggravation of symptoms.' No doubt many of Hahnemann's patients, given small doses of his new remedies, recovered. Some may have recovered more quickly than was to be expected from previous experience of the disorders from which they were suffering. The real problem here is that people do recover from diseases, and there is no reason to suppose that, because an alleged remedy has been given, the recovery was the result of the 'remedy'. Because A follows B there is no reason, without further evidence, to believe that B was the cause of A. There is, or should be, in such cases, a clear distinction between consecutiveness and causality. Hahnemann's failure to make this distinction was a pardonable error: after all, it was not until well into the twentieth century that it was realized that very special precautions had to be taken to ensure that *post hoc* (after this) was really *propter hoc* (on account of this), and not just pure coincidence.

What was to come next is harder to justify. Hahnemann came to believe that the smaller the dose of the 'remedy', the greater was its curative effect. The logic of this has defeated scientists ever since. To achieve a small dose, Hahnemann diluted his medicine, but he did not call this 'dilution': he called it 'potentiation', surely one of the all-time masterpieces of optimism.

His preferred dilution was the thirtieth 'potentiation'. So far as one can judge from his writings, Hahnemann used the term 'first centesimal dilution' to mean a dilution of 1 part in 100, and 'second centesimal dilution' as 1 in 10,000, and so on; thus a thirtieth centesimal dilution would be extremely dilute: so dilute that it would be unlikely that the 'solution' contained even one single molecule of the original substance. However pure the water he was using, it would have to contain considerably greater quantities of other impurities than the quantity of the drug that was present.

Although most of Hahnemann's professional colleagues at once saw the absurdity of his claims, many of them did not. No doubt from similar motives to his, many doctors accepted the theory of homoeopathy and the method caught on. Today, homoeopathy is flourishing. It is very big business. Every year, millions of people swallow millions of pounds worth of totally useless pills and feel better for it. There is even a homoeopathic hospital in London.

Once upon a time there was a rich and powerful emperor whose principal interest in life was fine clothes. One day he was visited by two men who told him that they could weave a cloth finer than any ever seen before. The emperor was entranced and urged them to get to work at once. So the two men set up a loom and began to go through the motions of weaving. But there was nothing in the loom. If anyone came in he or she was asked: 'Can't you see the wonderful, magical cloth? It's so beautiful! If you can't see it you must be really stupid.' So the magical cloth was woven and cut and a splendid suit of clothes was made. Before the suit was shown to the emperor the news was spread around the town that the cloth was too fine to be visible to people who were either stupid or unfit for their jobs. The two men respectfully clothed the emperor in this suit – which was made of material so fine that the emperor couldn't feel any weight. 'How handsome you look!' said one of the men. 'You must walk in a parade to show all the people.' The populace was informed of the emperor's new clothes and everyone understood that these were magical clothes. The emperor, stark naked, led a splendid parade through the streets of his capital city, strutting proudly along between the assembled crowds. Everyone cheered and expressed unbounded admiration for the magnificence of the emperor's clothes. Everyone, that is, except one little boy who turned to his mother and asked: 'Why doesn't that man have anything on?'

One wonders whether Hans Andersen had been reading the *Organon*.

> The human embryo arises from the sperm alone.
>
> *Aristotle (384–322 BC)*

A quantum leap to the wrong conclusion

It is common, these days, to hear a certain kind of person assert that since the development of quantum physics we can no longer rely on science to provide a basis for ordinary practical affairs. They point out that Heisenberg's uncertainty principle holds that it is impossible to determine simultaneously the position and velocity of a subatomic particle such as an electron. So how can we say anything reliable about the physical world? Some go further and suggest that quantum theory – which is now universally accepted – forces us to abandon all the laws of nature of which science has for so long been so proud. Some theologians have even claimed that quantum theory actually cuts away all the objections that science has seemed to raise to dogmatic fundamental religious belief.

Let's take a look, then, at some of the features of quantum physics that cause people to have these perceptions of modern science and see whether they are justified. All physical scientists now accept that, so far as subatomic particles are concerned, it is impossible simultaneously to determine the position and velocity of a particle. This is not a matter of the crudeness of instruments; it is a fact of nature. Even if the instruments were capable of making the two measurements with total accuracy, they could not do so simultaneously. Moreover, uncertainty in one measurement increases as the accuracy of the other measurement increases.

Uncertainty also applies to other pairs of quantities, for example energy and time. If we know the energy of a particle, we can't tell how long it is likely to remain in that particular energy state; and if we know how long it has had a particular level of energy, we can't tell what the energy level is. This fact has an astonishing implication. Particles can't exist without energy. But the uncertainty principle implies that particles can briefly come into existence in the absence of enough energy for them to exist. Such particles are called virtual particles and they soon disappear. They always occur in pairs – as a particle and an antiparticle.

When two sets of waves, say water waves, come together, a wave pattern, known as an interference pattern, is formed. This is quite simple to understand. If a peak coincides with a peak, a higher peak will be formed. If a trough coincides with a trough, a lower trough will be formed. And if a peak coincides with a trough, they will cancel each other and the surface of the water will be flat. The same process can be demonstrated with sound waves which, if in phase with each other produce a louder sound, and, if out of phase with each other, can produce silence.

Light passing through two narrow slits close together will also produce interference patterns of brightness and darkness which show up on a screen as bright stripes. It was the observation of this fact that caused the great physicist and polymath Thomas Young (1773–1829) to conclude that light was a wave phenomenon. Now, electrons are particles. They are negatively charged and can easily be made to move by attracting them with a positive charge (unlike charges attract each other). This is how a TV tube works. Electrons are 'boiled' off a heated filament at the back of the tube and attracted to the inside of the front of the tube by a high positive voltage. The inside of the tube is coated with a phosphor that glows brightly when electrons strike it. But if electrons are fired at a thin metal plate with two very narrow slits in it, in spite of being particles they, too, when they strike the fluorescing screen, produce the same kind of interference pattern! This is extraordinary and there is no explanation for it other than that electrons are both particles and waves.

But what does a quantum physicist actually mean when he or she uses the word 'particle'? Some think of a particle as a short length of wave of a limited number of cycles. The length of the wave segment is roughly equal to the size of the particle but, of course, there is no particular point along that wave segment that can, with certainty, be called the precise position of the particle. The physicist envisages a particle as an entity that is brought into existence as a result of the interference of many waves. This idea fits better with the overall concept of energy and matter being interchangeable – as well as with many physical phenomena – than the idea that a particle is a little solid ball.

When an electron on an atom is hit by a photon and changes its energy level, it doesn't *move* from one orbit to another. It just disappears from one and instantaneously appears in the other. When a body radiates energy, such as heat, or light, or X-rays, there is no way we can tell which particle (or photon) is going to be emitted next. The thing is entirely indeterminate. This being so, we can have no way of knowing what causes the emission of the particle – an instance of the failure of causality.

Problems such as these were very vexing to the physicists of the time, and caused no end of arguments. So Niels Bohr (1885–1962) decided to put together a kind of statement about the implications of quantum phenomena in the hope of getting a consensus. This has been accepted by many and, as a courtesy to the Dane Bohr, has been called the Copenhagen interpretation. The essence of this is that, in the subatomic world it is impossible to make any observations without changing what you are looking at. So it is impossible to know how anything happens. All we can know is the outcome, and we can never know that anything will happen for certain. All we can know is that there is a certain probability that something will happen. The outcome of every interaction depends on pure chance.

So, can the facts of quantum physics affect people in the real, familiar, macroscopic world – the world of gross objects composed of countless millions of atoms and their constituent particles? This brings us to the vexed question of Schrödinger's cat. This cat is entirely fictitious and was invented up by Erwin Schrödinger (1887–1961) to highlight a shortcoming of the Copenhagen interpretation. Schrödinger wanted to point out that the Copenhagen interpretation deals with the transition from an indeterminate quantum state to a definite macroscopic state in an unsatisfactory way; and devised the following thought experiment. A cat is sealed in a box that also contains a spot of very weakly radioactive material. This is so placed that if a particle is given off, it could hit a radiation detector and the resulting amplified signal would release some poison gas and kill the cat. The chances (probability) of a particle being released in a period of 1 minute are arbitrarily taken to be 50 : 50.

According to the Copenhagen interpretation, at the end of 1 minute it is completely meaningless for anyone outside the box to say that the cat is either alive or dead. It is not just that we don't know; the fact, according to the Copenhagen interpretation, is that the system has no definite state until it is observed. Until we open the box, the cat is neither alive nor dead. It exists in a 'superposed' state of alive/dead. By dreaming up this idea, Schrödinger appeared, quite effectively, to reduce the Copenhagen interpretation to absurdity. Einstein agreed with Schrödinger. 'God,' he said, 'does not play dice.' Yet, today, most physicists are willing to go along with this implication.

The facts of quantum physics are certainly remarkable and are such as to lead people to strange conclusions. But it is essential never to forget that these mysterious events are properties only of the quantum world. Anyone who extrapolates directly from the quantum world to the 'real' world of gross objects, goes right off the rails. Quantum theory is concerned with what happens to individual particles – particles actually smaller than atoms. Our physical world is made up of countless numbers of such particles. Although we cannot determine the position and speed of a single particle, what we can do is to predict with high accuracy the statistical effect of such enormous numbers of particles. In the statistical world of huge numbers, probabilities add up to reliable facts. Anyone who refused to accept the statistical basis for the structure of matter would have to feel rather precarious sitting on a chair. It is ridiculous to suggest that because we can't predict the position and velocity of any particle, we can't rely on matter to remain 'solid'.

As a matter of fact, quantum theory does a rather better job than classical physics in determining solidity. It doesn't even allow for the existence of a vacuum. All space, it appears, is filled with pairs of virtual particles – such as protons and antiprotons – that can spring into existence at any moment.

Are you sitting comfortably?

> The blood slays the bacillus without any help from the phagocytes.
>
> *George Bernard Shaw (1856–1950)*

The clinical trial ethical paradox

Most clinical trials compare one treatment with another to see which works better. The old and the new treatments are allotted randomly and neither the patients nor the doctors know, until the end of the trial, who has had which. Hence the term 'double-blind'. Very few trials would be mounted if the people concerned did not think that they had a worthwhile new treatment, and in most cases, they are likely to believe that the new treatment is better than the old. If they did not, they would hardly bother to try it out. Clinical trials are expensive and time-consuming.

But there is a surprising anomaly in all this. If there is any good reason to think that the new treatment is better than the old, the trial should not be conducted at all. This may sound paradoxical, but it is not. If there is real evidence of superiority of the new treatment, then it is unethical to encourage patients to take part in the trial because half of them will be getting what is believed to be an inferior treatment.

As the Helsinki Declaration of the World Health Organization states: 'The interests of science and society should never take precedence over the well-being of the subject. In any medical study, every patient – including those of a control group – should be assured of the best proven diagnostic and therapeutic method.' So if drug manufacturers and doctors are to behave as they should, there ought, strictly, to be far fewer clinical trials.

The position is even more delicate when, as is often the case, a new treatment is compared with a dummy, or placebo, treatment which is randomly and anonymously allocated. On the face of it, this might seem to be an excellent way to proceed, as it eliminates the element of suggestion. But is it? If a placebo is being compared with a treatment believed to be of some medical value, then half the participants in the trial are being denied treatment altogether.

Drug manufacturers, however ethical their aspirations, have to be in it for the money. If they don't make profits, the business fails and the executives get sacked. Costs are horrendous and failure to compete disastrous. In such an environment, is there any real likelihood that ethical counsels such as these will prevail? Not much.

There's something rotten in the state of medicine.

> Modern science kills God and takes his place on the vacant throne. Science is the sole legitimate arbiter of all relevant truth.
>
> *Vaclav Havel (1936–)*

The greatest scientific blunder?

The enormous emphasis science places on rationality has driven many scientists to a total rejection of religion. Today, there is an ongoing tussle between religious fundamentalists and scientists. Fundamentalists hold that all statements in the Bible are literally true, in particular those statements made at the beginning of the book of Genesis about the creation of the world. For this reason Western religious fundamentalists are often called 'Creationists'. In the Western world, the predominant religion is Christianity, but, to some extent, what follows applies to other religions also. The question we have to ask ourselves is this: are scientists who reject religion because they cannot accept the literal truth of the Bible behaving in a rational manner? Before we can answer this we must take a reasonably detailed look at the difficulties scientists have in accepting the content of the Bible.

The Bible is not, of course, the only religious book. To millions of highly religious people, it is either unknown or ignored or considered of secondary importance to their own holy books. Nearly all religions acquire an extensive documentation, produced either by those to whom God is believed to have spoken or by those claiming some special insight into the nature of God. Such books are written by people who have experienced what is called revelation or who have witnessed events believed to manifest the workings of God upon humanity. The criteria by which candidate documents are included in the corpus of holy script have generally been determined by scholars. Once so incorporated, such documents acquire a kind of authority which tends to perpetuate their inclusion.

Although the Bible is not unique as a religious document, it is particularly relevant in this context because the Western conflict between religion and science has been based very largely on the Bible. The first thing to be said is that the Bible, as accessible to nearly everyone today, is a translation. There is nothing unique about the Authorized Version with its 'biblical' language – which is actually the seventeenth-century language of the educated people who edited the version. This is only one of various English translations used today, and all of these are translations of translations. The original text of the Old Testament, in Hebrew and Aramaic, was the work of various unknown writers. Most of the New Testament was originally written in Greek and Aramaic. In the third century BC, Greek was the predominant language and scholars spent a hundred years translating the original Old Testament texts into Greek. As Christianity spread, there were further translations into Coptic, Latin, Ethiopian and Gothic. One Latin translation – that of St Jerome, completed in AD 405 became the 'official' version of Christianity for over 1000 years. It should also be remembered that, prior to the invention of printing, Bibles could be reproduced only by manual copying from one to another – a process certain to introduce errors, some of which would be perpetuated.

The first complete English translation was that of John Wycliffe (*c.*1329–84) and his committee, produced in 1382. This was revised in 1388. Between 1525 and 1535, the scholar William Tyndale (*c.*1494–1536) translated the New Testament and part of the Old Testament. This became the model for later English translations. At about the same time, Martin Luther (1483–1546), in Germany, finished his translation from Hebrew and Greek into German. This was the first produced in a fairly modern European language.

The first version to be recognizable today as a familiar Bible was the King James version – the 'Authorized Version' – produced in 1611, the work of 54 scholars. This version had a long life and is still preferred by many, but has been succeeded by numerous other versions, including the American Standard Version of 1901, the version of James Moffatt (1870–1944) of 1924, the translation of Ronald Knox (1888–1957) of 1949 widely used by Roman Catholics, the Revised Standard Version of 1952, the New English Bible of 1961, the Jerusalem Bible of 1966, the New American Bible of 1970, the Good News Bible of 1976 and the New International Version of 1978.

The book of Genesis provides an account of the creation of the world. In the first verse, dealing with the first day of creation, the text states: 'God divided the light from the darkness. And God called the light Day, and the darkness He called Night.' In the fourth verse, dealing with the fourth day, the text states: 'And God made two great lights; the greater light to rule the day, and the lesser light to rule the night: he made the stars also. And God set them in the firmament of the heaven to give light upon the earth.' Now, if this is literally true, we are already into a grave irrationality. If the sun and the moon were not created until the fourth day, there cannot have been light and darkness or morning and evening, as stated to occur in the earlier days.

A little later it states: 'So God created man in his own image; . . . male and female created he them. And God blessed them, and God said unto them, "Be fruitful and multiply, and replenish the earth, and subdue it . . ." ' Then there is the story of the 'fall' in the Garden of Eden, in which, after eating the forbidden fruit Adam and Eve 'knew that they were naked' and made clothes. The whole implication is that of an undesirable awareness of sexuality – without which it would not be possible to 'be fruitful and multiply'. If all this is literally true, God's command also implies the necessity for incest. If Adam and Eve were the only humans created, they could not otherwise be the source of all humanity. But incest is universally condemned as a sin and a crime.

The Genesis creation story is categorically contradicted by unequivocal scientific facts derived from geology and palaeontology. The fossil record is there. Darwinian evolution may not be an established scientific fact, but evolution certainly is. The only other possibility would be that when God created the world, he also created the fossil record. Irrationalities of this kind make it impossible for intellectually honest scientists to accept the biblical creation story as being literally true. But, again, that has nothing to do with whether or not scientists can be religious.

The Bible, being composed of many different works by different writers, contains a number of intrinsic contradictions. The God as described in the Old Testament, for instance, is very different from the God as described by Christ in the New Testament. It is unnecessary to go into details but the contrast between the two accounts could hardly be more striking. The God of the Old Testament is a reflection of the savagery of the times in which these books were written; that of the Gospels in the New Testament is a reflection of the thought of the greatest religious, social and political reformer of all time – Jesus Christ. They cannot both be literally true. No two contradictions can both be true. To claim that they can is to reduce reason to nonsense.

Then there is the question of biblical miracles. If the matter is to be considered rationally, it is necessary to enquire into the evidence we have for believing in miracles. The eminent Scottish philosopher and historian David Hume (1711–76) had a cogent comment on this point. Here is a short extract from his book *An Enquiry concerning Human Understanding*: 'No testimony is sufficient to establish a miracle, unless the testimony be of such a kind that its falsehood would be even more miraculous than the fact which it endeavours to establish. When anyone tells me that he saw a dead man restored to life, I immediately consider with myself whether it be more probable that this person should either deceive or be deceived, or that the fact which he relates should really have happened. I weigh the one miracle against the other; and according to the superiority which I discover, I pronounce my decision, and always reject the greater miracle.'

Questions such as that of the resurrection of bodies after death as promised in the Bible, and that of eternal Heaven or Hell, cannot even begin to be considered in a rational way. The very terms of the propositions are intrinsically irrational. If a person dies and the body is buried or burnt so that physical dissolution is complete, it is possible only by magical thinking to suggest that the prior body can be restored in living form. This is a fantastic suggestion that no scientist could ever subscribe to without abandoning reason.

Many such items in the Bible are obviously the product of the physical hardship, scientific ignorance and simplicity of thought current at the time these stories were written. Some were clearly rationalizations of the authors' inability to accept the fact of personal death. Poverty, hunger, heat, dust, dirt, and lack of comfort and luxury will, of course, induce dreams of a paradise of the kind described in religious books, and, in some cases, hatred of those who oppose or reject the views of the writer may have induced the classical concept of Hell. This latter mechanism has, of course, operated for centuries after the biblical texts were written.

Today, enlightened and educated members of the Church hierarchy are the last to insist on the literal truth of the Bible. Modern churchmen are well aware of its contradictions, inconsistencies and irrationalities. Most of them would agree that the opening verses of Genesis are now to be read as a metaphorical and poetical account of the creation. Whether they were believed by the author or authors – who, of course, did not have the benefit of modern scientific knowledge – is neither here nor there.

Religion is commonly, but not necessarily, associated with a formal code of morality. This is especially so in the major religions such as Christianity, Judaism, Islam, Hinduism and Buddhism. The degree of adherence to the ethical code of a religion is not, however, necessarily taken as an index of the spirituality of a person. In many cases, adherence to the theological rules and formal observances is considered more important. Christianity, in particular, has always been more tolerant of sinners than of heretics. The significance of this may be that, while human behaviour, ethical or otherwise, is a legitimate concern of science, spirituality is not. There is a quaint idea that unless a person adheres to a religious faith he or she has no basis for moral conduct. This is, of course, ridiculous. Non-religious people do not behave any worse than religious people. Some of them even believe that it is better to have rational reasons for good behaviour than to behave well for fear of punishment.

Scientists are as entitled as any other intelligent people to comment on, and be influenced by, the things that humans – mostly men – have done with religion. They are aware that, like all other major and important human activities, religion has become formalized into hierarchical organizations. The basis of the place in the hierarchy, from the humblest priest to the *Pontifex maximus*, ought, they suppose, to be the level of spirituality or the ability to inspire spirituality in others, but they may doubt whether this is always actually so. Scientists have good reason to suppose that in religious hierarchies the motives for seeking advancement are often much the same as those in secular hierarchies such as the learned professions, politics and business. It seems to many scientists unlikely that the ideals of the founders of religions are equalled by those who strive to rise in religious hierarchies.

Some scientists are deterred by ritual and the use of prescribed liturgy. Some are put off by archaic language and are aware that sometimes its meaning is unknown to the lay participants. They know that many people, for instance, oppose the substitution of contemporary language for that of the Authorized Version of the Bible, for ritual purposes. Few can remain unperturbed by the uses religion has been put to, and by the gross abuses that have been perpetrated in its name. Atrocities such as those of the Inquisition with its inhuman torture and burning of millions; the persecution and burning of witches in the name of the Church; and wars, such as the Crusades against the 'infidel', have clearly shown many scientists how disastrous the abandonment of reason and the adoption of superstition and magic can be.

Many scientists would hold that inability to accept as literally true some of the content of the Bible, is not a criticism of God. It is an implicit criticism of the claims of men as to the status of the Bible as a whole. Unfortunately, some religious people insist that rejection of any statement in the Bible – which they insist is the 'word of God' – is blasphemy and an attack on religion. Many scientists have been so disgusted by the kind of irrationalities described above that they have, on that account alone, rejected religion totally. Many others, however, have seen the error of this and hold that although the Bible reflects many human qualities, bad as well as good, and that these qualities include ignorance, prejudice and error, it is an immensely important work.

Attempts to write or talk about God immediately raise the problem that we have no intellectual equipment to handle transcendental matters. The only thing we can do is to reduce them to human terms and to express them in terms of human emotion. Religious belief usually involves a sense of the sacred or holy – a sense of a reality that transcends science and the whole natural world and that is considered of ultimate value. The human mind cannot easily conceive of or use such a concept in the abstract, so it is usually personalized as a God. The Bible explicitly states that God made man 'in his own image', but it is more obviously true that man has made God in his own image.

In conceiving God, even the most sophisticated have great difficulty in avoiding this human-shaped, or anthropomorphic, image. Whatever the nature of God may be, if God created the universe, his image cannot be that of a human except in the most superficial and unimportant sense. The transcendental being, however visualized, is credited with supernatural powers, and religious faith commonly incorporates the conviction that these powers can be employed in the interests of, or to the disadvantage of, the believer. In many cases, the God is believed to be in a permanent personal relationship with the believer.

In this shadowy borderland between the familiar physical world and the world of the spirit, rational mental processes soften and dissolve. Often without our realizing it we drift into the use of propositions and logic that may not be applicable. Some people who have thought about this have concluded that 'conventional' reason does not apply in such matters. They will say: 'Oh, yes, when we're talking about ordinary everyday affairs, reason is appropriate; but when we are dealing with subjects like the possibility of life after death, or the existence of God, or the creation of the world, ordinary logic no longer applies. You can't use the same rules when considering such matters as you do when you are dealing with science.' One possible implication of this is that there is a form of logic that *can* be used in considering matters that transcend ordinary experience. Another is that no meaningful discussion of such matters is possible.

Let's see where these ideas take us. We'll start by considering where we get our concept of reason. One thing is certain – we are not born with it. Our ideas of cause and effect, for instance are derived from our experience. We see certain things happening and notice that they have certain consequences. Some seem obvious. Step off a precipice and you fall. Others are less obvious and require repeated experience before we begin to get the idea. If an approaching car is a mile down the road, it is usually safe to cross. Very rarely, it is not. Eventually we come to refine our understanding that things happen in a certain way or in a certain order, never otherwise. It is one of the major functions of science to produce a systematic and formal description of these things. In time, experience makes many of these observations so obvious that we begin to call them logical. We don't need a physics textbook to tell us that a man who is shot, killed and cremated does not turn up alive and well a month later. To say that he can is what we call an irrational statement and we reject it out of hand.

The point of all this is that our ideas of reason are derived from observations of the real world. None of us has ever had any experience of a total environment or system other than the world as we all know it. This is not metaphysics, just plain common sense. As children, we all engage in non-rational thought, sometimes called magical thinking. This is because children have not yet had time to experience enough of how the universe really is, and so to build up the ability to think rationally. Children will commonly, for instance, blame themselves for the death of a parent because they know they have been naughty. The 'logical' link between their behaviour and the death of the parent is an example of pre-rational magical thinking. To varying degrees, this particular kind of irrationality persists into adult life. All of us indulge in it to some extent.

Rationality – conformity to the patterns of the world as we know it – is important to us, so important that we feel distinctly uncomfortable when we come across instances of gross irrationality. We dislike the idea of insanity because it features irrational thought that may lead to irrational action. A schizophrenic person may commit 'justifiable' murder under the profoundly irrational conviction that the victim is about to destroy the world by starting an atomic chain reaction. Rationality is also, of course, important to us because, since rationality is a reflection of the way the world is, we can get into real trouble if we choose to ignore it. People will often allow the strength of their appetites or instincts to overrule their reason, and the results are nearly always at least unfortunate and often disastrous.

Science is the supreme expression of rationality. Fully established scientific facts have been subjected to a process of scrutiny and testing that is far more rigorous than, for instance, those applied to political opinions or economic 'principles'. When scientific research throws up seemingly irrational facts – such as some of those of quantum mechanics – this is not a breakdown of rationality. It is a reflection of the fact that there has been an important gap in our prior knowledge or experience, so that our rationality, which was based on that limited experience, has to be extended to include the new facts.

At present, the understanding of quantum mechanics is at a stage of development at which there are features of it that no one can understand. This is not a breakdown of reason, and it is certainly not a cause for throwing reason out of the window. That there are things in quantum mechanics that we can't understand is no reason to say that they are wrong or that we don't believe them. That would be irrational. However seemingly incredible they may be, these are observable facts and must be accepted, and to fit them into a consistent scientific scheme we need reason and further research.

So, for the scientist, the application of strict rationality is a vital tool. Scientists do not believe in magic. They do not believe in miracles. If a blind person with structurally normal eyes goes to a faith-healer and is completely restored to normal vision, the scientist has no problem in believing it. The scientist knows that there are mechanisms that can cause blindness in the absence of organic disease of the eyes or the brain, and that such blindness can be reversed. But if a blind person with organic disease that has turned his optic nerves into fibrous tissue states that he believes he can be restored to vision by a faith-healer, the scientist will be distressed and possibly outraged.

So, in the light of all this, can there be a system of logic that is valid for matters that transcend human experience? The difficulty is that we can see nothing on which such a logic can be based. If the only conceivable basis for reason is properly tested experience, then matters of which we have no experience cannot be a basis for a logic. The question of applying 'conventional' logic to transcendental matters is equally difficult. It is common to hear religious people talking of God in exactly the same terms as they talk of everyday affairs. They use logical arguments to discuss theology, with no regard to at least the possibility that these arguments may be meaningless. When they use terms like 'the soul', for instance, they are referring to an entity of which none of them has any real knowledge.

But, for all these difficulties, the scientific community contains many devotedly religious people who recognize the sharp limitations of science in tackling the problems of a suffering humanity and in answering the perennial questions of humankind. Some scientists who see that the nature of God is unknowable – in the sense of the word 'knowable' familiar to science – and who reject most of theological thought and argument as meaningless, will tell you that, rationally, these are reasons to accept rather than to reject religion. They are intimations of the way in which religion differs fundamentally from science.

To many scientists, as to others, religious faith is not simply a passive belief in matters which cannot be scientifically demonstrated; it is a positive decision to accept certain hypotheses and to act upon them, like a leap in the dark, in the expectation that the outcome will be good. People of intense religious beliefs will usually assert that the process works.

There is nothing irrational about that.

Bibliography

Angelo Jr, J. A. *The Dictionary of Space Technology*. Frederick Muller, 1982.

Appleyard, B. *Understanding the Present*. Picador, 1992.

Appleyard, R. *A Tribute to Michael Faraday*. Constable, 1931.

Asimov, I. *Biographical Encyclopedia of Science and Technology*. Doubleday & Co., 1964.

Bailey, G. *Galileo's Children*. Little, Brown & Co., 1990.

Barnes-Svarney (ed.) *The New York Public Library Science Desk Reference*. MacMillan, New York, 1995.

Berridge, V. and Edwards, G. *Opium and the People*. Yale University Press, 1987.

Boltz, C. L. *Ernest Rutherford*. Heron Books, 1970.

Bradlaugh, C. *Humanity's Gain from Unbelief*. Watts & Co., 1929.

Brock, W. H. *The Fontana History of Chemistry*, 1992.

Calhoun, D. R. (ed.) *Britannica Book of the Year*. Encyclopædia Britannica, 1987–97.

Curie, E. *Madame Curie*. Heinemann, 1938.

DeBeer G. *Charles Darwin*. Nelson, 1963.

Dewdney, A. K. *Yes, We have No Neutrons*. Wiley, 1997.

Dingle, H. (ed.) *A Century of Science*. Hutchinson's Scientific and Technical Publications, 1951.

Dobbs, B. J. T. *The Foundations of Newton's Alchemy*. Cambridge University Press, 1975.

Draper, J. W. *History of the Conflict Between Religion and Science*. Kegan Paul, 1890.

Dunsheath, P. *A History of Electrical Engineering*. Faber & Faber, 1962.

Einstein, A. *Relativity*. Wings Books, New York, 1961.

Emiliani, C. *The Scientific Companion*. Wiley, 1988.

Eve, A. S. *Rutherford*. Cambridge University Press, 1939.

Fabricius, J. *Alchemy*. Diamond Books, 1989.

Ferris T. *The Whole Shebang*. Weidenfeld & Nicolson, 1997.

Ferris, T. (ed.) *The World Treasury of Physics, Astronomy and Mathematics.* Little, Brown & Co., 1991.

Forrest, D. W. *Francis Galton: The Life and Work of a Victorian Genius.* Paul Elek, 1974.

Gell-Mann, M. *The Quark and the Jaguar.* Little, Brown & Co., 1994.

Gillespie, C. C. *Genesis and Geology.* Harper & Row, 1951.

Haeckel, E. *The Riddle of the Universe.* Watts & Co., 1929.

Hall, A. Rupert (ed.) *The Rise of Modern Science.* Collins, 1962.

Hawking, S. W. *A Brief History of Time.* Bantam Press, 1988.

Hawking, S. W. *Black Holes and Baby Universes.* Bantam Press, 1993.

Heilbron, J. *Elements of Early Modern Physics.* University of California Press, 1982.

Heisenberg, W. *Physics and Philosophy.* Penguin Books, 1990.

Hey, T. and Walters, P. *The Quantum Universe.* Cambridge University Press, 1987.

Hoffmann, B. *The Strange Story of the Quantum.* Pelican, 1965.

Hoffmann, B. *Einstein.* Paladin, 1973.

Holmyard, E. *Alchemy.* Penguin Books, 1968.

Huxley, L. *Life and Letters of Thomas Henry Huxley.* Macmillan & Co., 1900.

Irvine, W. *Apes, Angels and Victorians. A Joint Biography of Darwin and Huxley.* Weidenfeld & Nicolson, 1955.

Jungk, R. *Brighter Than a Thousand Suns.* Pelican Books, 1956.

Kingsford, P. W. *Electrical Engineers and Workers.* Edward Arnold, 1968.

Krauss, L. *Fear of Physics.* Jonathan Cape, 1994.

Lodge, O. *Pioneers of Science.* Macmillan & Co., 1905.

Maclaren, M. *The Rise of the Electrical Industry during the Nineteenth Century.* Princeton University Press, 1943.

McKenzie, A. E. E. *The Major Achievements of Science.* Cambridge University Press, 1960.

Medawar, P. *The Threat and the Glory.* HarperCollins, 1990.

Medawar, P. *Pluto's Republic.* Oxford University Press, 1984.

Metchnikoff, O. *Life of Elie Metchnikoff.* Constable, 1921.

Metchnikoff, E. *The Nature of Man.* Heinemann, 1906.

Millar, D., Millar I., Millar, J. and Millar, M. *The Cambridge Dictionary of Scientists.* Cambridge University Press, 1996.

Milsted, David. *They Got it Wrong.* Guinness Publishing, 1995.

Moulton, F. R. and Schifferes, J. J. (eds) *The Autobiography of Science.* John Murray, 1963.

Muir, H. (ed.) *Larousse Dictionary of Scientists.* Larousse, 1994.

Murray, R. H. *Science and Scientists in the Nineteenth Century.* The Sheldon Press, 1925.

Overbye, D. *Lonely Hearts of the Cosmos.* Macmillan, 1991.

Parry, M. (ed.) *Chambers Biographical Dictionary.* Chambers, 1997.

Penrose, R. *The Emperor's New Mind.* Vintage, 1989.

Pole, W. *The Life of Sir William Siemens.* John Murray, 1888.

Polkinghorne, J. *Rochester Roundabout: The Story of High Energy Physics.* Longman, 1989.

Popper, K. R. *The Logic of Scientific Discovery.* Hutchinson, 1972.

Porter, R. (ed.) *The Hutchinson Dictionary of Scientific Biography.* Helicon, 1994.

Priestley, J. Of dephlogisticated air. In: *Nature and Nature's Laws.* Macmillan, 1970.

Rapoport, Y. *The Doctors' Plot.* Fourth Estate, 1991.

Ronan, C. A. *The Cambridge Illustrated History of the World's Science.* Book Club Associates, 1983.

Routledge, R. *Discoveries and Inventions of the 19th Century.* Bracken Books, 1989.

Russell, B. *ABC of Relativity.* George Allen & Unwin, 1971.

Sakharov, A. *Memoirs.* Hutchinson, 1990.

Sarton, G. (ed.) *A History of Science.* Oxford University Press, 1959.

Schrödinger, E. *What is Life?* Cambridge University Press, 1967.

Singer, C. *A Short History of Scientific Ideas.* Oxford University Press, 1959.

Stannard, R. *Science and Wonders.* Faber & Faber, 1996.

Sutherland, S. *Irrationality. The Enemy Within.* Constable, 1992.

Sutton, C. (ed.) *Building the Universe.* Blackwell and New Scientist, 1985.

Szydło, Z. *Water Which Does Not Wet Hands: The Alchemy of Michael Sendivogius.* Warsaw, 1994.

Thompson, C. J. S. *The Lure and Romance of Alchemy.* Harrap & Co., 1932.

Tolstoy, I. *The Knowledge and the Power.* Canongate, 1990.

Vallery-Radot, R. *The Life of Pasteur.* Garden City Publishing Co.

Webster, R. *Why Freud was Wrong.* HarperCollins, 1995.

Williamson, R. *The Making of Physicists.* Adam Hilger, 1987.

Wolpert, L. *The Unnatural Nature of Science.* Faber & Faber, 1992.

Wolpert, L. and Richards, A. *A Passion for Science.* Oxford University Press, 1988.

Wood, A. *Thomas Young, Natural Philosopher.* Cambridge University Press, 1954.

Bibliography

Youngson, R. M. *The Guinness Encyclopedia of Science*. Guinness Publishing, 1994.
Youngson, R. M. *The Guinness Encyclopedia of the Human Being*. Guinness Publishing, 1994.

Index

Index